Chemistry: A Very Short Introduction

Very Short Introductions available now:

ACCOUNTING Christopher Nobes
ADVERTISING Winston Fletcher
AFRICAN AMERICAN RELIGION
    Eddie S. Glaude Jr.
AFRICAN HISTORY John Parker and
    Richard Rathbone
AFRICAN RELIGIONS
    Jacob K. Olupona
AGNOSTICISM Robin Le Poidevin
ALEXANDER THE GREAT
    Hugh Bowden
AMERICAN HISTORY Paul S. Boyer
AMERICAN IMMIGRATION
    David A. Gerber
AMERICAN LEGAL HISTORY
    G. Edward White
AMERICAN POLITICAL HISTORY
    Donald Critchlow
AMERICAN POLITICAL PARTIES
    AND ELECTIONS L. Sandy Maisel
AMERICAN POLITICS
    Richard M. Valelly
THE AMERICAN PRESIDENCY
    Charles O. Jones
AMERICAN SLAVERY
    Heather Andrea Williams
THE AMERICAN WEST Stephen Aron
ANAESTHESIA Aidan O'Donnell
ANARCHISM Colin Ward
ANCIENT EGYPT Ian Shaw
ANCIENT EGYPTIAN ART AND
    ARCHITECTURE Christina Riggs
ANCIENT GREECE Paul Cartledge
THE ANCIENT NEAR EAST
    Amanda H. Podany

ANCIENT PHILOSOPHY Julia Annas
ANCIENT WARFARE
    Harry Sidebottom
ANGELS David Albert Jones
ANGLICANISM Mark Chapman
THE ANGLO-SAXON AGE John Blair
THE ANIMAL KINGDOM
    Peter Holland
ANIMAL RIGHTS David DeGrazia
THE ANTARCTIC Klaus Dodds
ANTISEMITISM Steven Beller
ANXIETY Daniel Freeman and
    Jason Freeman
THE APOCRYPHAL GOSPELS
    Paul Foster
ARCHAEOLOGY Paul Bahn
ARCHITECTURE Andrew Ballantyne
ARISTOCRACY William Doyle
ARISTOTLE Jonathan Barnes
ART HISTORY Dana Arnold
ART THEORY Cynthia Freeland
ASTROBIOLOGY David C. Catling
ATHEISM Julian Baggini
AUGUSTINE Henry Chadwick
AUSTRALIA Kenneth Morgan
AUTISM Uta Frith
THE AVANT GARDE
    David Cottington
THE AZTECS David Carrasco
BACTERIA Sebastian G. B. Amyes
BARTHES Jonathan Culler
THE BEATS David Sterritt
BEAUTY Roger Scruton
BESTSELLERS John Sutherland
THE BIBLE John Riches

BIBLICAL ARCHAEOLOGY
    Eric H. Cline
BIOGRAPHY Hermione Lee
THE BLUES Elijah Wald
THE BOOK OF MORMON
    Terryl Givens
BORDERS Alexander C. Diener and
    Joshua Hagen
THE BRAIN Michael O'Shea
THE BRITISH CONSTITUTION
    Martin Loughlin
THE BRITISH EMPIRE Ashley Jackson
BRITISH POLITICS Anthony Wright
BUDDHA Michael Carrithers
BUDDHISM Damien Keown
BUDDHIST ETHICS Damien Keown
CANCER Nicholas James
CAPITALISM James Fulcher
CATHOLICISM Gerald O'Collins
CAUSATION Stephen Mumford and
    Rani Lill Anjum
THE CELL Terence Allen and
    Graham Cowling
THE CELTS Barry Cunliffe
CHAOS Leonard Smith
CHEMISTRY Peter Atkins
CHILDREN'S LITERATURE
    Kimberley Reynolds
CHILD PSYCHOLOGY Usha Goswami
CHINESE LITERATURE Sabina Knight
CHOICE THEORY Michael Allingham
CHRISTIAN ART Beth Williamson
CHRISTIAN ETHICS D. Stephen Long
CHRISTIANITY Linda Woodhead
CITIZENSHIP Richard Bellamy
CIVIL ENGINEERING David Muir Wood
CLASSICAL LITERATURE William Allan
CLASSICAL MYTHOLOGY
    Helen Morales
CLASSICS Mary Beard and
    John Henderson
CLAUSEWITZ Michael Howard
CLIMATE Mark Maslin
THE COLD WAR Robert McMahon
COLONIAL AMERICA Alan Taylor
COLONIAL LATIN AMERICAN
    LITERATURE Rolena Adorno
COMEDY Matthew Bevis
COMMUNISM Leslie Holmes
COMPLEXITY John H. Holland
THE COMPUTER Darrel Ince

CONFUCIANISM Daniel K. Gardner
THE CONQUISTADORS
    Matthew Restall and
    Felipe Fernández-Armesto
CONSCIENCE Paul Strohm
CONSCIOUSNESS Susan Blackmore
CONTEMPORARY ART
    Julian Stallabrass
CONTEMPORARY FICTION
    Robert Eaglestone
CONTINENTAL PHILOSOPHY
    Simon Critchley
CORAL REEFS Charles Sheppard
CORPORATE SOCIAL RESPONSIBILITY
    Jeremy Moon
COSMOLOGY Peter Coles
CRITICAL THEORY
    Stephen Eric Bronner
THE CRUSADES Christopher Tyerman
CRYPTOGRAPHY Fred Piper and
    Sean Murphy
THE CULTURAL REVOLUTION
    Richard Curt Kraus
DADA AND SURREALISM
    David Hopkins
DARWIN Jonathan Howard
THE DEAD SEA SCROLLS Timothy Lim
DEMOCRACY Bernard Crick
DERRIDA Simon Glendinning
DESCARTES Tom Sorell
DESERTS Nick Middleton
DESIGN John Heskett
DEVELOPMENTAL BIOLOGY
    Lewis Wolpert
THE DEVIL Darren Oldridge
DIASPORA Kevin Kenny
DICTIONARIES Lynda Mugglestone
DINOSAURS David Norman
DIPLOMACY Joseph M. Siracusa
DOCUMENTARY FILM
    Patricia Aufderheide
DREAMING J. Allan Hobson
DRUGS Leslie Iversen
DRUIDS Barry Cunliffe
EARLY MUSIC Thomas Forrest Kelly
THE EARTH Martin Redfern
ECONOMICS Partha Dasgupta
EDUCATION Gary Thomas
EGYPTIAN MYTH Geraldine Pinch
EIGHTEENTH-CENTURY BRITAIN
    Paul Langford

THE ELEMENTS Philip Ball
EMOTION Dylan Evans
EMPIRE Stephen Howe
ENGELS Terrell Carver
ENGINEERING David Blockley
ENGLISH LITERATURE Jonathan Bate
ENTREPRENEURSHIP
    Paul Westhead and Mike Wright
ENVIRONMENTAL ECONOMICS
    Stephen Smith
EPIDEMIOLOGY Rodolfo Saracci
ETHICS Simon Blackburn
ETHNOMUSICOLOGY Timothy Rice
THE ETRUSCANS Christopher Smith
THE EUROPEAN UNION
    John Pinder and Simon Usherwood
EVOLUTION Brian and
    Deborah Charlesworth
EXISTENTIALISM Thomas Flynn
EXPLORATION Stewart A. Weaver
THE EYE Michael Land
FAMILY LAW Jonathan Herring
FASCISM Kevin Passmore
FASHION Rebecca Arnold
FEMINISM Margaret Walters
FILM Michael Wood
FILM MUSIC Kathryn Kalinak
THE FIRST WORLD WAR
    Michael Howard
FOLK MUSIC Mark Slobin
FOOD John Krebs
FORENSIC PSYCHOLOGY
    David Canter
FORENSIC SCIENCE Jim Fraser
FOSSILS Keith Thomson
FOUCAULT Gary Gutting
FRACTALS Kenneth Falconer
FREE SPEECH Nigel Warburton
FREE WILL Thomas Pink
FRENCH LITERATURE John D. Lyons
THE FRENCH REVOLUTION
    William Doyle
FREUD Anthony Storr
FUNDAMENTALISM Malise Ruthven
GALAXIES John Gribbin
GALILEO Stillman Drake
GAME THEORY Ken Binmore
GANDHI Bhikhu Parekh
GENES Jonathan Slack
GENIUS Andrew Robinson

GEOGRAPHY John Matthews and
    David Herbert
GEOPOLITICS Klaus Dodds
GERMAN LITERATURE Nicholas Boyle
GERMAN PHILOSOPHY Andrew Bowie
GLOBAL CATASTROPHES Bill McGuire
GLOBAL ECONOMIC HISTORY
    Robert C. Allen
GLOBAL WARMING Mark Maslin
GLOBALIZATION Manfred Steger
GOD John Bowker
THE GOTHIC Nick Groom
GOVERNANCE Mark Bevir
THE GREAT DEPRESSION AND THE
    NEW DEAL Eric Rauchway
HABERMAS James Gordon Finlayson
HAPPINESS Daniel M. Haybron
HEGEL Peter Singer
HEIDEGGER Michael Inwood
HERODOTUS Jennifer T. Roberts
HIEROGLYPHS Penelope Wilson
HINDUISM Kim Knott
HISTORY John H. Arnold
THE HISTORY OF ASTRONOMY
    Michael Hoskin
THE HISTORY OF LIFE
    Michael Benton
THE HISTORY OF MATHEMATICS
    Jacqueline Stedall
THE HISTORY OF MEDICINE
    William Bynum
THE HISTORY OF TIME
    Leofranc Holford-Strevens
HIV/AIDS Alan Whiteside
HOBBES Richard Tuck
HORMONES Martin Luck
HUMAN EVOLUTION Bernard Wood
HUMAN RIGHTS Andrew Clapham
HUMANISM Stephen Law
HUME A. J. Ayer
HUMOUR Noël Carroll
THE ICE AGE Jamie Woodward
IDEOLOGY Michael Freeden
INDIAN PHILOSOPHY Sue Hamilton
INFORMATION Luciano Floridi
INNOVATION Mark Dodgson and
    David Gann
INTELLIGENCE Ian J. Deary
INTERNATIONAL MIGRATION
    Khalid Koser

INTERNATIONAL RELATIONS
    Paul Wilkinson
INTERNATIONAL SECURITY
    Christopher S. Browning
IRAN Ali M. Ansari
ISLAM Malise Ruthven
ISLAMIC HISTORY Adam Silverstein
ITALIAN LITERATURE
    Peter Hainsworth and David Robey
JESUS Richard Bauckham
JOURNALISM Ian Hargreaves
JUDAISM Norman Solomon
JUNG Anthony Stevens
KABBALAH Joseph Dan
KAFKA Ritchie Robertson
KANT Roger Scruton
KEYNES Robert Skidelsky
KIERKEGAARD Patrick Gardiner
KNOWLEDGE Jennifer Nagel
THE KORAN Michael Cook
LANDSCAPE ARCHITECTURE
    Ian H. Thompson
LANDSCAPES AND
    GEOMORPHOLOGY
    Andrew Goudie and Heather Viles
LANGUAGES Stephen R. Anderson
LATE ANTIQUITY Gillian Clark
LAW Raymond Wacks
THE LAWS OF THERMODYNAMICS
    Peter Atkins
LEADERSHIP Keith Grint
LINCOLN Allen C. Guelzo
LINGUISTICS Peter Matthews
LITERARY THEORY Jonathan Culler
LOCKE John Dunn
LOGIC Graham Priest
LOVE Ronald de Sousa
MACHIAVELLI Quentin Skinner
MADNESS Andrew Scull
MAGIC Owen Davies
MAGNA CARTA Nicholas Vincent
MAGNETISM Stephen Blundell
MALTHUS Donald Winch
MANAGEMENT John Hendry
MAO Delia Davin
MARINE BIOLOGY Philip V. Mladenov
THE MARQUIS DE SADE John Phillips
MARTIN LUTHER Scott H. Hendrix
MARTYRDOM Jolyon Mitchell
MARX Peter Singer

MATERIALS Christopher Hall
MATHEMATICS Timothy Gowers
THE MEANING OF LIFE
    Terry Eagleton
MEDICAL ETHICS Tony Hope
MEDICAL LAW Charles Foster
MEDIEVAL BRITAIN John Gillingham
    and Ralph A. Griffiths
MEMORY Jonathan K. Foster
METAPHYSICS Stephen Mumford
MICHAEL FARADAY
    Frank A. J. L. James
MICROBIOLOGY Nicholas P. Money
MICROECONOMICS Avinash Dixit
THE MIDDLE AGES Miri Rubin
MINERALS David Vaughan
MODERN ART David Cottington
MODERN CHINA Rana Mitter
MODERN FRANCE
    Vanessa R. Schwartz
MODERN IRELAND Senia Pašeta
MODERN JAPAN
    Christopher Goto-Jones
MODERN LATIN AMERICAN
    LITERATURE
    -Roberto González Echevarría
MODERN WAR Richard English
MODERNISM Christopher Butler
MOLECULES Philip Ball
THE MONGOLS Morris Rossabi
MORMONISM
    Richard Lyman Bushman
MUHAMMAD Jonathan A. C. Brown
MULTICULTURALISM Ali Rattansi
MUSIC Nicholas Cook
MYTH Robert A. Segal
THE NAPOLEONIC WARS
    Mike Rapport
NATIONALISM Steven Grosby
NELSON MANDELA Elleke Boehmer
NEOLIBERALISM Manfred Steger and
    Ravi Roy
NETWORKS Guido Caldarelli and
    Michele Catanzaro
THE NEW TESTAMENT
    Luke Timothy Johnson
THE NEW TESTAMENT AS
    LITERATURE Kyle Keefer
NEWTON Robert Iliffe
NIETZSCHE Michael Tanner

NINETEENTH-CENTURY BRITAIN
Christopher Harvie and
H. C. G. Matthew
THE NORMAN CONQUEST
George Garnett
NORTH AMERICAN INDIANS
Theda Perdue and Michael D. Green
NORTHERN IRELAND
Marc Mulholland
NOTHING Frank Close
NUCLEAR POWER Maxwell Irvine
NUCLEAR WEAPONS
Joseph M. Siracusa
NUMBERS Peter M. Higgins
NUTRITION David A. Bender
OBJECTIVITY Stephen Gaukroger
THE OLD TESTAMENT
Michael D. Coogan
THE ORCHESTRA D. Kern Holoman
ORGANIZATIONS Mary Jo Hatch
PAGANISM Owen Davies
THE PALESTINIAN-ISRAELI
CONFLICT Martin Bunton
PARTICLE PHYSICS Frank Close
PAUL E. P. Sanders
PEACE Oliver P. Richmond
PENTECOSTALISM William K. Kay
THE PERIODIC TABLE Eric R. Scerri
PHILOSOPHY Edward Craig
PHILOSOPHY OF LAW
Raymond Wacks
PHILOSOPHY OF SCIENCE
Samir Okasha
PHOTOGRAPHY Steve Edwards
PHYSICAL CHEMISTRY Peter Atkins
PLAGUE Paul Slack
PLANETS David A. Rothery
PLANTS Timothy Walker
PLATO Julia Annas
POLITICAL PHILOSOPHY David Miller
POLITICS Kenneth Minogue
POSTCOLONIALISM Robert Young
POSTMODERNISM Christopher Butler
POSTSTRUCTURALISM
Catherine Belsey
PREHISTORY Chris Gosden
PRESOCRATIC PHILOSOPHY
Catherine Osborne
PRIVACY Raymond Wacks
PROBABILITY John Haigh

PROGRESSIVISM Walter Nugent
PROTESTANTISM Mark A. Noll
PSYCHIATRY Tom Burns
PSYCHOLOGY Gillian Butler and
Freda McManus
PSYCHOTHERAPY Tom Burns and
Eva Burns-Lundgren
PURITANISM Francis J. Bremer
THE QUAKERS Pink Dandelion
QUANTUM THEORY
John Polkinghorne
RACISM Ali Rattansi
RADIOACTIVITY Claudio Tuniz
RASTAFARI Ennis B. Edmonds
THE REAGAN REVOLUTION Gil Troy
REALITY Jan Westerhoff
THE REFORMATION Peter Marshall
RELATIVITY Russell Stannard
RELIGION IN AMERICA Timothy Beal
THE RENAISSANCE Jerry Brotton
RENAISSANCE ART
Geraldine A. Johnson
REVOLUTIONS Jack A. Goldstone
RHETORIC Richard Toye
RISK Baruch Fischhoff and John Kadvany
RIVERS Nick Middleton
ROBOTICS Alan Winfield
ROMAN BRITAIN Peter Salway
THE ROMAN EMPIRE
Christopher Kelly
THE ROMAN REPUBLIC
David M. Gwynn
ROMANTICISM Michael Ferber
ROUSSEAU Robert Wokler
RUSSELL A. C. Grayling
RUSSIAN HISTORY Geoffrey Hosking
RUSSIAN LITERATURE Catriona Kelly
THE RUSSIAN REVOLUTION
S. A. Smith
SCHIZOPHRENIA Chris Frith and
Eve Johnstone
SCHOPENHAUER Christopher Janaway
SCIENCE AND RELIGION
Thomas Dixon
SCIENCE FICTION David Seed
THE SCIENTIFIC REVOLUTION
Lawrence M. Principe
SCOTLAND Rab Houston
SEXUALITY Véronique Mottier
SHAKESPEARE Germaine Greer

SIKHISM Eleanor Nesbitt
THE SILK ROAD James A. Millward
SLEEP Steven W. Lockley and
    Russell G. Foster
SOCIAL AND CULTURAL
    ANTHROPOLOGY
    John Monaghan and Peter Just
SOCIALISM Michael Newman
SOCIOLINGUISTICS John Edwards
SOCIOLOGY Steve Bruce
SOCRATES C. C. W. Taylor
THE SOVIET UNION Stephen Lovell
THE SPANISH CIVIL WAR
    Helen Graham
SPANISH LITERATURE Jo Labanyi
SPINOZA Roger Scruton
SPIRITUALITY Philip Sheldrake
SPORT Mike Cronin
STARS Andrew King
STATISTICS David J. Hand
STEM CELLS Jonathan Slack
STRUCTURAL ENGINEERING
    David Blockley
STUART BRITAIN John Morrill
SUPERCONDUCTIVITY
    Stephen Blundell
SYMMETRY Ian Stewart
TEETH Peter S. Ungar
TERRORISM Charles Townshend
THEATRE Marvin Carlson

THEOLOGY David F. Ford
THOMAS AQUINAS Fergus Kerr
THOUGHT Tim Bayne
TIBETAN BUDDHISM
    Matthew T. Kapstein
TOCQUEVILLE Harvey C. Mansfield
TRAGEDY Adrian Poole
THE TROJAN WAR Eric H. Cline
TRUST Katherine Hawley
THE TUDORS John Guy
TWENTIETH-CENTURY BRITAIN
    Kenneth O. Morgan
THE UNITED NATIONS
    Jussi M. Hanhimäki
THE U.S. CONGRESS
    Donald A. Ritchie
THE U.S. SUPREME COURT
    Linda Greenhouse
UTOPIANISM Lyman Tower Sargent
THE VIKINGS Julian Richards
VIRUSES Dorothy H. Crawford
WITCHCRAFT Malcolm Gaskill
WITTGENSTEIN A. C. Grayling
WORK Stephen Fineman
WORLD MUSIC Philip Bohlman
THE WORLD TRADE
    ORGANIZATION Amrita Narlikar
WORLD WAR II Gerhard L. Weinberg
WRITING AND SCRIPT
    Andrew Robinson

Available soon:

PLATE TECTONICS Peter Molnar
DANTE Peter Hainsworth and
    David Robey

HUMAN ANATOMY Leslie Klenerman
ANCIENT ASSYRIA Karen Radner
TAXATION Stephen Smith

For more information visit our website

www.oup.com/vsi/

Peter Atkins

CHEMISTRY

A Very Short Introduction

OXFORD
UNIVERSITY PRESS

Great Clarendon Street, Oxford, OX2 6DP,
United Kingdom

Oxford University Press is a department of the University of Oxford.
It furthers the University's objective of excellence in research, scholarship,
and education by publishing worldwide. Oxford is a registered trade mark of
Oxford University Press in the UK and in certain other countries

First published in hardback as *What is Chemistry?* 2013
First published as a Very Short Introduction 2015

The moral rights of the author have been asserted

First edition published in 2015

Impression: 1

Published in the United States of America by Oxford University Press
198 Madison Avenue, New York, NY 10016, United States of America

British Library Cataloguing in Publication Data
Data available

Library of Congress Control Number: 2014948110

ISBN 978-0-19-968397-0

Printed in Great Britain by
Ashford Colour Press Ltd, Gosport, Hampshire

Contents

Preface xiii

1 Its origins, scope, and organization 1

2 Its principles: atoms and molecules 13

3 Its principles: energy and entropy 27

4 Its reactions 38

5 Its techniques 53

6 Its achievements 66

7 Its future 86

Periodic table 95

Glossary 97

Further reading 101

Index 103

Preface

I hope to open your eyes and show you a fascinating, intellectually and economically important world, that of chemistry. Chemistry, I have to admit, has an unhappy reputation. People remember it from their schooldays as a subject that was largely incomprehensible, fact-rich but understanding-poor, smelly, and so far removed from the real world of events and pleasures that there seemed little point in coming to terms with its grubby concepts, spells, recipes, and rules. In later life that unhappy reputation is often rendered unhappier still by an awareness of the environmental impact of nasty chemicals escaping into the wild and bringing disaster to softly green clover-clad bucolic meadows that were home to the glowing poppy and the dancing butterfly, rendering into inhospitable mud the banks where the wild thyme once grew, generating toxic sludge and noxious slime where limpid streams had rippled, replacing air fragrant with aeolian delight with pungency, and generally messing things up.

I want to change all that. I want to encourage you to look anew at chemistry, through modern unprejudiced eyes, with those memories and attitudes swept away and replaced by comprehension and appreciation. I want to show you the world through a chemist's eyes, to understand its central concepts, and see how a chemist contributes not only to our material comfort but also to human culture. I want to explain how chemists think and how what they reveal about matter—all forms of matter, from rocks to humans—adds pleasure to our perception of the world. I want to show you how chemists take one form of matter, perhaps sucked or dug from the ground or plucked from the skies, and turn it into another form, perhaps to clothe us, feed us, or comfort us.

I want to share with you the thought that chemistry provides the infrastructure of the modern world. There is hardly an item of everyday life that is not furnished by it or based on the materials it has created. Take away chemistry and its functional arm the chemical industry and you take away the metals and other materials of construction, the semiconductors of computation and communication, the fuels of heating, power generation, and transport, the fabrics of clothing and furnishings, and the artificial pigments of our blazingly colourful world. Take away its contributions to agriculture and you let people die, for the industry provides the fertilizers and pesticides that enable dwindling lands to support rising populations. Take away its pharmaceutical wing and you allow pain through the elimination of anaesthetics and deny people the prospect of recovery by the elimination of medicines. Imagine a world where there are no products of chemistry (including pure water): you are back before the Bronze Age, into the Stone Age: no metals, no fuels except wood, no fabrics except pelts, no medicines except herbs, no methods of computation except with your fingers, and very little food.

Advances in technology demand the availability of materials with new and sophisticated properties, be it better electrical, magnetic, optical, or mechanical properties or just greater purity. Advances in the maintenance of human health that can reduce the demand on the physical infrastructure of hospitals and their sophisticated, expensive equipment depend on the discovery and manufacture of better, more sophisticated medicines. There will be no advances in the generation, deployment, and conservation of energy without chemistry to provide its material infrastructure.

It goes without saying, however, that the extraordinary difference between raw nature and what chemistry transforms it into to enhance and extend our lives comes at a price, and it is that price that disconcerts us and is rightly the basis of our apprehension of chemistry's environmental impact. At its crudest, the products of chemistry enhance our ability to kill and maim, for weaponry is improved when new explosives and other agents are perfected. Often of more permanent and vocal concern is the undeniable environmental impact of what is produced and the processes of production. Chemistry puts into societies' hands the ability by governmental choice to wage war more effectively, through commercial pressures to produce artefacts more aggressively, and through personal choice to squander more profligately and thereby harm our unique and irreplaceable ecosystem.

I shall confront that concern in these pages, for it has been a corollary of progress in chemical manufacturing and the presence not only of its products but also of its manufacturing waste in the environment. It is important, though, to bear in mind a rounded picture of chemistry, not a single black facet. Without chemistry life would be nasty, brutish, and short. With chemistry, it can be comfortable, entertaining, and well fed. Transport can be efficient; clothes alluring. Lives can be longer. Without ignoring the dark and negative side of chemistry, I shall encourage you to appreciate the illuminating and positive side, too.

There is another dimension to all these contributions: understanding. Chemistry provides insight into the heart of matter by showing how things are. A chemist can look on a rose and understand why it is red and look on a leaf and understand why it is green. A chemist can look on glass and understand why it is brittle and look on a fabric and understand why it is supple. The glories of Nature, of course, can be experienced without this inner knowledge, just as music can be enjoyed without analysis; but the insight that chemistry brings into the properties of matter, in all its forms, can be brought to bear if the moment is apt, and deeper enjoyment thereby achieved. I seek to share some of this insight with you and show that even a little chemistry will add to your daily pleasure.

That, in broad terms, is the journey I shall take you on. I shall try to dislodge you from your half-remembered, perhaps unpleasant memories of your early encounter with chemistry. You will not have a degree in chemistry when you have read through these chapters, for chemistry is deep as well as wide, it is quantitative as well as qualitative, it is subtle as well as superficial. You will, however, I hope, appreciate its structure, its core concepts, and its contributions to culture, pleasure, economy, and the world.

In conclusion, I would like to thank Professor David Phillips, Imperial College, for a number of helpful remarks.

Peter Atkins
Oxford, 2014

Chapter 1
Its origins, scope, and organization

Greed. Greed inspired humanity to embark on an extraordinary journey that touches everyone today. The particular variety of greed I have in mind was jointly the quest for immortality and the attainment of unbounded riches. The supposed route to both was the manipulation of matter to provide elixirs to overcome bodily ills for the realization of immortality and recipes for the conversion of more or less anything resembling gold—either in colour, as in urine and sand, or heft, as in lead—into gold itself. Neither aim was ever achieved, but the ceaseless tinkering with matter by the alchemists provided them with a considerable familiarity with it and provided the compost, often literally, from which a real science—chemistry—was to emerge.

The principal instrument of the transition from alchemy to chemistry was the balance. The ability to weigh things precisely put into humanity's hands the potential to attach numbers to matter. The significance of that achievement should not go by unremarked, for it is in fact quite extraordinary that meaningful numbers can be attached to air, water, gold, and every other kind of matter. Thus, through the attachment of numbers, the study of matter and the transformations that it can undergo (the current scope of chemistry) was brought into the domain of the physical sciences, where qualitative concepts can be rendered quantitatively and tested rigorously against the theories that surround and illuminate them.

Weighing matter before and after it had undergone transformation from one substance to another led to the principal concept that underlies all explanations in chemistry: the *atom*. The concept of the 'atom' had floated around groundlessly in human consciousness for over two millennia, ever since the ancient Greeks had speculated, without an iota of evidence, for some kind of ultimate indivisible particulate graininess of the world. Their speculation became grounded in science in the hands of John Dalton (1766–1844), who through the analysis of the weights of substances before and after reaction drew the conclusion that the elements, the fundamental building blocks of matter, are composed of unchangeable atoms, and that track could be kept of them as one substance changed into another by the simple expedient of weighing.

Atoms are now the currency of chemistry. Almost every explanation in chemistry refers to them, either as individuals or strung together in the combinations we call *molecules*. Atoms are the constituents of all matter: everything you can see and touch is built of atoms. As small as they are it is quite wrong to say that they are invisible to the naked eye. Look at a tree: you are seeing atoms. Look at a chair: you are seeing atoms. Look at this page: you are seeing atoms (even if this page is on a screen). Touch your face: you are touching atoms. Touch a fabric: you are touching atoms. Of course, an individual atom is too small to see: but matter is built from battalions of them, and the swarming battalions are visible to the naked eye as the substances that surround us. Later, however, in Chapter 5, I shall explain how chemists can now even see images of *individual* atoms.

There are just over 100 different types of atom. Quite what I mean by 'type' I shall explain in Chapter 2 when together we look inside them and identify their differing internal structures that render them distinct. Each different type of atom corresponds to a different element. Thus, just as there are the elements hydrogen, carbon, iron, and so on, so there are hydrogen atoms, carbon atoms, iron atoms, and so on, all the way up to the most recently discovered element, which in 2013 is the wholly useless and exceedingly short-lived 114th element, livermorium. (To be precise: it is element 116, but two that precede it await discovery). The key idea in chemistry is that when one substance changes into another, the atoms themselves do not change: they simply exchange partners or enter into new arrangements. Chemistry is all about divorce and remarriage.

Although 'atom' means uncuttable, atoms are cuttable. Even armchair speculation leads to that conclusion, for the existence of different types of atom implies the possession of different structures, so with sufficient ingenuity it is likely that an atom can be blasted apart and the so-called *subatomic particles* from which it is formed identified. Experiment confirms this speculation, and we shall see something of the interior of atoms and thereby the origins of their different personalities in Chapter 2. It is here that chemistry draws most strongly on physics, for physicists unravelled the structures of atoms and chemists use this information to account for the molecules they form and the reactions they undergo.

That last remark hints at the scope of chemistry. It implies that to understand chemistry it is necessary to import concepts from physics. That is indeed the case, and chemistry draws heavily on numerous concepts developed by physicists (in return, we chemists provide the matter for them to conjure with). Among all this trade there are two hugely important imports, one relating to the behaviour of individual atoms and their subatomic components and the other relating to bulk, that is tangibly large versions of matter, such as a jug of water or a block of iron. More technically, these are the *microscopic* and *macroscopic* worlds, respectively.

The crucial import from physics to account for the properties of the microscopic world of individual atoms and molecules is *quantum mechanics*. Although much of chemistry was developed during the 19th century, there was little understanding of why some things occurred and others did not. At that time, Isaac Newton's 'classical mechanics', the mathematical procedures for accounting for the motion of bodies, was king, for it was so successful at accounting for the orbits of planets and the flight of balls, and there was the expectation that when planets and balls were slimmed down to atoms, explanations of chemistry would be found and Newton's domain would encompass chemistry too. Newton's fruitless focus on alchemical manipulation was perhaps a sign that he thought so too. However, at the end of the 19th century and early in the 20th it was found that this slimming down of planets and balls to atoms resulted in the complete failure of classical mechanics: even the concepts on which Newton's mechanics was based crumbled when applied to atoms and their constituents. Such are the dangers of uncircumspect extrapolation.

Then, early in the 20th century, around 1927, a new mechanics was born that has proved to be hugely successful for explaining how atoms and subatomic particles go about their business. To this day the theory, quantum mechanics, has not been superseded in predictive power and numerical precision. That it remains largely incomprehensible is admittedly an irksome deficiency, but in due course I shall do my best to distil from it what is necessary for understanding the behaviour of atoms and hence the whole of chemistry. We shall see that when chemists stir and boil their fluids, they are coaxing atoms to behave according to the weird laws of quantum mechanics.

The other crucial import from physics, in this case to account for the properties of the macroscopic world of bulk matter, is *thermodynamics*. Thermodynamics is the science of energy and the transformations it can undergo. It arose in large part through the Victorian era's dependence on the steam engine for driving societies forward both literally and economically, but soon proved to be a key part of the fabric of chemistry. The material fabric of our subject is atoms, but the changes they undergo are under the control and impetus of energy. We shall see that not only is energy released when a fuel burns—an obvious, useful but primitive aspect of the involvement of energy with chemistry—but also that it governs how atoms behave in general, what structures they can form, what changes in organization they can undergo, and at what rate those changes can occur. Energy also, in a subtle way, turns out to be the driving power of chemistry in the sense that reactions are impelled forward by it in a manner that I shall explain in Chapter 3. Because energy is so intimately coiled into the very structure of chemistry, it should not be surprising that thermodynamics plays a role despite its engineering origins.

Whereas chemistry reaches down into physics for its explanations (and through physics further down into mathematics for its quantitative formulation), it reaches upwards into biology for many of its most extraordinary applications. That should not be surprising, for biology is merely an elaboration of chemistry. Before biologists explode in indignation at that remark, which might seem akin to claiming that sociology is an elaboration of particle physics, let me be precise. Organisms are built from atoms and molecules, and those structures are explained by chemistry. Organisms function, that is, are alive, by virtue of the complex network of reactions taking place within them, and those reactions are explained by chemistry. Organisms reproduce by making use of molecular structures and reactions, which are both a part of chemistry. Organisms respond to their environment, such as through olfaction and vision, by changes in molecular structure, and thus those responses—all our five or so senses—are elaborations of chemistry. Even that hypermacroscopic phenomenon, evolution and the origin of species, can be regarded as an elaborate working out of the consequences of the Second Law of thermodynamics, and is thus an aspect of chemistry. Some organisms, I have in mind principally human beings, cogitate on the nature of the world, and the mental processes that underlie and are manifest as these cogitations are due to elaborate networks of chemical reactions. Thus, biology is indeed an elaboration of chemistry. I shall not press the view, whatever I actually think, that all matters of interest to biologists, such as animal behaviour in general, are also merely elaborated chemistry, but confine myself to the assertion that all the structures, responses, and processes of organisms are chemical. Chemistry thus pervades biology, and has contributed immeasurably to our understanding of organisms.

We socially elaborate organisms, we humans, build things. We fabricate artefacts. We mine the stones of the Earth, pump the fluids from the deep, and harvest the gases of the skies and aim to turn all this raw material into whatever we desire. The conversion of those raw materials into substances that can be moulded, hammered, spun, glued together, eaten, or simply burned, is a part of chemistry. Chemists might step aside and allow moulders to mould, hammerers to hammer, shapers to shape, and in general fabricators to fabricate, to create the final artefact, but it is they who have provided the raw material, the infrastructure of our modern technological society, and have contributed hugely thereby to world economies and the deportment of individuals and nations.

As I have emphasized in the Preface, there are, of course, speckles and blotches of black amid all this light. Chemistry has certainly contributed to mankind's ability to maim and kill, and it would be inappropriate in this survey of what chemistry is to sweep under the carpet of its pages its provision of explosives, of nerve gases, and its accidental and intentional impositions on our fragile environment. I shall confront these issues later, but at this stage—to emphasize the importance of personal judgement—I invite you to eliminate all the contributions of chemistry to the modern world, which will take you back to the painful, dangerous, uncomfortable, aspirationally restricted era of the Stone Age, and to ask yourself then whether the current darkness outweighs the light.

The divisions of chemistry

The scope of chemistry, then, is so enormous that my introduction to it, and the subject itself, would wallow amorphously like a stranded spineless whale without the imposition of some kind of structure. Chemists have drifted into a structure that helps them to carry out their activities, congregate in like-minded assemblies, and develop their procedures much like individual states develop their policies and economies. Unlike most states, the boundaries are blurred, and often striking advances are made where two cultures overlap. That is especially the case when the subject is as mature as chemistry currently is, where each domain of activity is thoroughly explored and inspiration might most fruitfully come, just like in art, at fertile overlapping boundaries and at the frontiers where chemistry overlaps other disciplines.

For our purposes, and to understand the general structure of chemistry for the sake of this introduction, it is helpful to appreciate its division into various branches and to see in broad terms their concerns. The divisions of chemistry still pervade university departments, courses, and the journals where discoveries are reported, and so a description of them is still an important component of a visitor's guidebook. But be warned: frontiers both intellectual and departmental are melting.

The broadest, most important and conventional, and still widely observed division of chemistry is into its physical, organic, and inorganic branches.

Physical chemistry lies at the interface of physics and chemistry (hence its name) and deals with the principles of chemistry which, as we have seen, consist largely of quantum mechanics for explaining the structures of atoms and molecules and thermodynamics for assessing the role and deployment of energy. It is also concerned with the rates at which reactions take place, both at the macroscopic level and the microscopic. In the latter it seeks to follow the intimate lives of individual molecules as they are ripped apart and then reconstituted as different substances in reactions. A major activity of physical chemistry is its contribution to the interpretation of investigative techniques, particularly 'spectroscopy'.

As we shall see in Chapter 5, spectroscopy uses various kinds of light to bring information from within molecules into the eyes, increasingly the synthetic eyes, of the observer. Such is the current sophistication of these techniques that physical chemists must bring all their armoury, particularly quantum mechanics, to bear on the interpretation of the data. Indeed, so blurred are the activities of chemists and physics in this domain that the name physical chemistry often elides into *chemical physics* for some who study the behaviour of individual molecules with an approach that lies close to a physicist's.

Organic chemistry is the part of chemistry that is concerned with the compounds of carbon. That one element can command a whole division is a testament to carbon's pregnant mediocrity. Carbon lies at the midpoint of the Periodic Table, the chemist's map of chemical properties of the elements, and is largely indifferent to the liaisons it enters into. In particular, it is content to bond to itself. As a result of its mild and unaggressive character, it is able to form chains and rings of startling complexity. Startling complexity is exactly what organisms need if they are to be regarded as being alive, and thus the compounds of carbon are the structural and reactive infrastructure of life. So extensive are the compounds of carbon, currently numbering in the millions, that it is not surprising that a whole branch of chemistry has evolved for their study and has developed special techniques, systems of nomenclature, and attitudes.

Why 'organic'? Such is the intricacy of the molecules to which carbon contributes (except for a few outliers, like simple carbon dioxide), that it was originally thought that only Nature could form them. That is, according to this 'vitalist' view, they are the products of organisms. The beginning of the end of vitalism was in 1828, when it was shown that a simple mineral could be converted into a characteristic 'organic' compound (namely urea). Although dispute raged for some time, since then the 'organic' of organic chemistry has been an archaism; but convenient archaisms are hard to dislodge and the term survives but now means nothing more than 'a compound of carbon'.

That leaves the rest of the elements, the hundred or so elements other than carbon. Their study is the domain of *inorganic chemistry*. As might be suspected about a branch of a subject that deals with over 100 elements with widely differing personalities, inorganic chemistry is a vital yet sprawling field of study. The sprawl is partly contained by the adoption of various subdivisions of the subject. A major subdivision is *solid-state chemistry*, where the object of study is inorganic solids, such as the materials that act as superconductors and the semiconductors that have made universal computation feasible. It is hard to resist the analogy between inorganic chemistry and a hundred-piece orchestra, with the chemist conductor–composer drawing out symphonies of combinations by ordering the instruments accordingly.

Carbon is not secure from an inorganic chemist's Periodic-Table-scanning eyes. Some of the simpler compounds of carbon, such as the carbon dioxide that I have already mentioned, the killer gas carbon monoxide, and chalk and limestone that form our landscapes, are readily released by organic chemists from their domain as being of little interest to them and by convention are regarded as inorganic. On the frontier between the divisions, though, lie compounds that are intricate assemblies of carbon atoms yet include atoms of various metals. A number of these compounds are essential catalysts in the chemical industry; some are crucial to the functioning of organisms. Here lies the interdivisional field of *organometallic chemistry*, which at its best represents a highly fruitful collaboration between organic and inorganic chemists.

Chemistry's overlap with other disciplines

Such are the three principal divisions of chemistry. That list by no means exhausts all the ways in which chemists carve up their subject for better digestion, but all the others draw technique, concept, and inspiration from these three in various proportions and spice their mixture with aspects of other subjects. It would be a sizeable undertaking to list them all, but it is appropriate to be aware of the most common of them.

Analytical chemistry is the modern descendant of the age-old quest for finding out what is there. What is present in a mineral? Might there be gold or is it hafnium? What is present in crude oil? What is present in it other than the raw hydrocarbons, and which hydrocarbons? What is that compound you made? Can you deduce the arrangements of its atoms? These are all questions that analytical chemists might try to answer. Although test-tubes, flasks, and retorts still figure in their approaches, many of their investigations are now carried out in sophisticated machines, some of which use spectroscopy and others techniques developed by inorganic and physical chemists. I explore these techniques in Chapter 4. Stemming from analytical chemistry is *forensic chemistry*, in which the techniques of analytical chemistry are used for legal purposes, to track down or exonerate suspects, and to analyse the scenes of crimes.

Biochemistry is organic chemistry's back-donation to biology, sometimes with a dash of inorganic chemistry thrown in. It is concerned wholly with the structures and reactions that constitute living things, resolving the metabolic pathways that turn food into action (including that action confined to the brain: thought). Organisms are still a hugely important reservoir of organic molecules, for Nature has had billions of years to explore structural niches, and biochemists play a central role in both discovering what is there and working out how it was made under the control of the worker-bees of the body, the proteins we call enzymes. One anthropocentric but important concern about the extinction of species is that it wipes out sources of intricate molecules that have taken millions of years to emerge.

The name of *industrial chemistry* speaks for itself. Here chemist meets engineer, and reactions established in test-tubes and their kin are scaled up to enormous size and rendered fit to contribute to commerce. Industrial chemists contribute enormously to the economy and to trade between nations. In the United Kingdom alone, chemicals contribute 20 per cent to the gross domestic product, and in the United States over 96 per cent of all manufactured goods are directly touched by chemistry. Such figures relating to manufactured chemicals are almost literally not to be sniffed at. A principal concern of current industrial chemistry is *green chemistry*, the intention being to minimize waste, thereby enhancing economy, and to minimize impact on the environment, which enhances acceptability and sustainability.

The contribution of chemistry to other disciplines

Chemistry owes many debts to the subjects that surround it in the intellectual landscape, but they owe debts to chemistry too.

Physics owes debts, particularly in the field of electronics and increasingly of photonics (the use of light instead of electrons to convey information and manipulate data). Chemists create the semiconductors without which computation would be confined to the industrial scale from which it first emerged. They also formulate the glasses used in optical fibres, without which the transfer of information would be hobbled.

Biology owes an enormous debt to chemistry, especially since the emergence of *molecular biology*, springing largely from the identification of the structure of DNA and its interpretation as the carrier of genetic information from generation to generation. It is almost no exaggeration to say that biology became a part of the physical sciences once the chemical component of its principal characteristic, reproduction, had been identified. Molecular biology is really a version of chemistry, and the current maturity of chemistry has enabled biology to become as lively as it has never been before. The collaboration of biology and chemistry that we call *medicinal chemistry* is one of chemistry's great and unarguably acceptable contributions to society.

Society owes chemistry another huge debt too, for as I have said in the Preface, it deploys the material contributions of chemistry everywhere, in medicine, agriculture, communication, transport, and all forms of construction, fabrication, and decoration. We personally also owe a debt to chemistry, for as I claimed there too it gives us each an inner eye to enjoy the world.

All this stems from an understanding of chemistry, which I shall now start to unfold.

Chapter 2
Its principles: atoms and molecules

Central to any discussion of chemistry is the Periodic Table, that masterpiece of organization formulated in the 19th century principally by Dmitri Mendeleev (1834–1907) and its basis understood in the 20th century once the structures of atoms had been explained. That the table is important is confirmed by its ubiquity: it hangs on laboratory and lecture room walls and is printed in every introductory chemistry textbook. There is a version towards the end of this book. Its importance, though, should not be overstated. Working chemists do not gaze at it each morning for inspiration or refer to it frequently during the day. Certainly they have it in mind, for its real importance is that it summarizes relationships between the elements and plays a crucial role in organizing information about them. Perhaps its most important role is in the teaching of chemistry, for instead of being confronted with the daunting task of learning the properties of a hundred elements, it enables their properties to be inferred from their location in the table and trends in properties to be identified and easily remembered. Indeed, Mendeleev was led to formulate his table as he prepared to write a textbook of introductory chemistry.

The Periodic Table portrays an extraordinary feature of matter: that the elements are related to one another. We are now so familiar with the table that that feature is easily forgotten. But imagine yourself in an era before the table had been formulated.

Then you would have known of the gas oxygen and the yellow solid sulfur, and would almost certainly not have dreamt that there could be any relation between them. You would have known of the largely inert gas nitrogen and the incandescent solid phosphorus, and would not have conceived that they were related. And what about red copper, lustrous silver, and glowing gold? A family? Surely not! How, indeed, is it even possible for different forms of matter to be brothers or cousins? Even the concept of family relationships between different substances was hardly conceivable.

The Periodic Table, though, reveals that the elements are indeed related to one another. Oxygen and sulfur are cousins and stand next to each other in the table; so are nitrogen and phosphorus; copper, silver, and gold are members of the same family and lie together. Their very different appearances are superficial differences, for when the reactions they take part in and the molecules they form are investigated, it turns out that there are deep similarities between these relatives. Those similarities stem from the structures of their atoms, and to understand them it is to these atoms that we must now turn.

The structure of atoms

It was perhaps a little disconcerting for me to mention in Chapter 1 that to understand the structures of atoms it would be necessary to turn to quantum mechanics and all its incomprehensibilities. However, I did also mention that I would distil from that extraordinary theory only concepts and information that concern us. With that restriction in mind, it turns out that atoms have a rather simple structure and that it is quite easy to understand relationships between the elements and to understand, as this account unfolds, why some combinations of atoms are allowed and others not.

The basic structure of an atom consists of a nucleus surrounded by a cloud of electrons. This is the 'nuclear atom', the model of an atom first identified by Ernest Rutherford (1871–1937) in 1911. The nucleus is positively charged, the electrons are negatively charged, and it is the attraction between these opposite charges that is responsible for the existence and survival of the atom. As is well known, atoms are very small: there are over a million carbon atoms in (the printed version of) the full stop at the end of this sentence. A nucleus is even smaller: if an atom were enlarged to the size of a football stadium, the nucleus would be the size of a fly at its centre.

I shall start at the centre of the atom and work out. A nucleus consists of two types of subatomic particle: protons and neutrons. As suggested by the p and n in their names, protons are positively charged and neutrons are electrically neutral. Apart from that, they are very similar, with almost the same mass. They are tightly gripped together in the nucleus, and it requires a major effort—something like a nuclear explosion—to shake them loose. In most of chemistry, with its relatively puny releases of energy, the nucleus remains unchanged and is a passive but important participant in the processes going on around it in test-tubes, beakers, and flasks.

The number of protons in the nucleus determines the chemical identity of the atom. Thus, an atom of hydrogen has one proton, an atom of helium has two, an atom of carbon has six, nitrogen seven, oxygen eight, and so on up to livermorium, with 116. The number of protons in the nucleus is called the *atomic number* of the element. At once, we arrive at the first extraordinary feature of elements: they can be put in order according to their atomic number. No longer are elements a random jumble. They lie in a definite sequence: hydrogen, helium, ... livermorium. Moreover, because the atomic number can be used as a kind of roll-call, chemists and physicists know that they have identified the elements for every atomic number up to 116 apart from (in 2014) 113 and 115. They know that none is missing except those two and whatever lies beyond 116.

The neutrons are just passengers in this roll-call. A nucleus has about the same number of neutrons as protons in the nucleus, and that number can vary slightly. As the number of neutrons does not affect the atomic number, the same element can have atoms of slightly different numbers of neutrons and therefore different masses. These different versions of the atoms of the same element are called *isotopes* because they live in the same place (*isos* = 'equal' + *topos* = 'place') in the Periodic Table. Thus, hydrogen has three isotopes: hydrogen itself (one proton, no neutrons), deuterium (one proton, one neutron), and tritium (one proton, two neutrons). The first of these isotopes of hydrogen is by far the most abundant; the nucleus of tritium barely holds together and is 'radioactive', emitting radiation as it falls apart after a few years (its 'half-life' is 12.3 years). Deuterium is 'heavy hydrogen', with each atom weighing about twice that of ordinary hydrogen. In combination with oxygen, it forms 'heavy water', which, because deuterium atoms are heavier than hydrogen atoms, is about 10 per cent heavier than ordinary water.

The atomic number, the number of protons and hence the positive charge of the nucleus, determines the number of electrons that surround it. An *electron* has the same magnitude of electric charge as a proton, but opposite in sign. Therefore, for an atom to be electrically neutral the number of electrons outside the nucleus must be the same as the number of protons inside the nucleus. That is, the number of electrons is equal to the atomic number. Thus, hydrogen (atomic number 1) has one electron, carbon (atomic number 6) has six electrons, and so on, up to livermorium with its 116 electrons. Electrons are much lighter than protons and neutrons (by a factor of nearly 2,000), so their presence barely affects the mass of an atom. They have a profound effect on the chemical and physical properties of the element, and almost all chemistry can be traced to their behaviour.

Chemists have little interest in nuclei except for their role in determining how many electrons surround them. There is one exception, the very special individual case of the nucleus of a hydrogen atom, a single proton. I shall explain its special role in Chapter 4.

As I have mentioned, all chemical reactions leave nuclei intact. In other words, chemical reactions do not change the identities of elements. At a blow, we can see why the alchemists' desperate search for means of converting lead (element 82, with 82 protons in its nucleus) into gold (element 79, with 79 protons in its nucleus) was doomed to failure: heating, stirring, banging, and stamping in frustration could not extract the tightly bound three protons from the nucleus that was necessary for 'transmutation', the conversion of one element into another. Transmutation can occur, but that, the result of *nuclear* reactions, is the domain of nuclear energy and nuclear physics. Chemists have a vital role to play in dealing with the consequences of nuclear processes, especially in preparing nuclear fuel and dealing with nuclear waste, but *chemical* reactions leave all nuclei intact, and at this stage only chemical reactions are our concern.

Electrons in atoms

My focus now turns to the hugely important properties of the electron clouds that surround a nucleus. I need to make more precise the nature and structure of those clouds, for they are not just regions of swirling mist.

Electrons surround the nucleus in layers, rather like real clouds lying above each other, but encircling the entire atom. The concept of an electron being a 'cloud' needs a quick word of explanation. The cloud is really a cloud of probability: where it is dense, the electron is likely to be found; where it is sparse the electron is unlikely to be found.

The laws of quantum mechanics ordain that up to two electrons surround the nucleus in the lowest layer, up to a further eight in the next surrounding layer, and then a further 18 in the next layer. We don't need to go beyond that, but a similar pattern with variations continues indefinitely as the number of electrons grows. This pattern means that in hydrogen a single electron surrounds the nucleus. In carbon, with its six electrons, two electrons form the lowest-level cloud and four more form surrounding clouds in the outer layer. You could think of atoms as nuclei surrounded by onion-like layers of clouds, each inner layer being completed before the next layer begins. Why there are these characteristic numbers (2, 8, 18...) for successive layers need not concern us, but is fully understood in terms of quantum mechanics.

You are now face-to-face with the explanation of the structure of the Periodic Table and the familial relationships of the elements. Keep an eye on the table at the end of this book and start at hydrogen with its single electron. Go to helium, with its two electrons. Now the first cloud layer is full and, simultaneously, we find ourselves on the far right of the table. The next electron, needed by lithium, has to become a cloud in the next surrounding layer. Stepping across the table as electrons are added, passing carbon, nitrogen, and oxygen on the way, we complete the layer at neon, another gas, like helium. The next added electron must start the next cloud layer, and brings us to sodium, on the far left of the table, an element strongly resembling lithium in the row above it, both with a single electron outside completed cores of clouds.

Everything should now be clear: the layout of the table represents the filling of the cloud layers, with one electron present in the layer on the left of the table and the layer completed on the right. For technical reasons that are fully understood, but which would be a distraction here, the order in which cloud layers are completed gets a little muddled after the first two rows of the table, and although the lengths of the rows are the numbers we have already seen, namely 2, 8, 18..., and can be discerned, they lie in a funny but understood order (the pattern of the Periodic Table is 2, 8, 8, 18, 18...).

The crucial point is that elements that lie beneath each other have very similar patterns of cloud coverage. That is the origin of family relationships: oxygen and its cousin sulfur in the row below have the same pattern of clouds, it is just that sulfur's final six electrons lie in a higher level than oxygen's final six. Likewise phosphorus's final five electrons lie at a higher-level layer than nitrogen's final five in the row above it.

It is often said that atoms are mostly empty space. That simply isn't true. The cloudlike distributions of electrons fill the whole of space around the tiny fly-in-a-stadium-sized nucleus. Admittedly the cloud is very thin in parts; but it is there and all-pervasive. The assertion that an atom is mostly empty space springs from the outmoded view that electrons are like tiny pointlike planets whizzing round the nucleus at great distances from it, with lots of emptiness in between. Quantum mechanics replaces that figure with the cloudlike distributions that I have described, clouds that, although greatly attenuated in parts, fill all space.

How atoms form bonds

The principal concern of chemistry is not so much with individual atoms but the compounds that they form by entering into a variety of liaisons with one another. There are literally millions of such liaisons that have been identified and many more that we know exist but have not been identified and named. The richness of our environment is due to this huge collection of compounds, and chemists spend most of their hours building new combinations of atoms or tearing compounds down to see how they are built. To do this effectively, they need to understand how atoms link together and what controls the links, the *chemical bonds*, that they can form to one another.

What holds atoms together to form identifiable compounds, such as water, salt, methane, and DNA? Can there be any combination of atoms, or are there reasons for Nature's restraint which chemists despite all their meddling cannot circumvent? Why is there variety in the world of substances, but apparently not random variety? These questions can be inverted: why don't all the atoms of the universe just clump together in one huge solid mass?

The answer to all these questions lies in those layers of clouds. Broadly speaking, there are energy advantages in an atom acquiring a complete cloud layer. It can do that in a variety of ways. One is to shed electrons from the outermost layer. This it is likely to do if there are not many electrons in that layer to begin with, which means that it is more likely to happen with atoms of elements on the left of the Periodic Table, at the beginning of each new row and each new layer of cloud. Alternatively, if it already has a lot of electrons in its outermost layer, then it might gain electrons from somewhere and so complete its layer. That is likely to happen if the layer is almost full, which is the case for atoms of elements on the right of the Periodic Table towards the right-hand end of a row. There is another way to complete their layers: atoms could share electrons from each other's outermost layer. That might happen when one atom is reluctant to release an electron fully because there is no energy advantage in it. That subtle mediocrity carbon forms most of its extraordinary liaisons this way.

As we have seen, atoms are electrically neutral, with the total negative charge of all its electrons matching and cancelling the total positive charge of all the protons in the nucleus. When an atom gains or loses an electron, the balance of charges is upset and the atom becomes an *ion*. An ion is simply an electrically charged atom; it is so called because it will move in response to an electric field, and 'ion' is the Greek word for going. An atom that has gained one or more electrons is negatively charged and is called an *anion*. One that has lost one or more electrons is positively charged and is called a *cation* (pronounced 'cat ion'). The 'an' and 'cat' prefixes are from the Greek words for 'up' and 'down' and reflect the fact that oppositely charged ions move in opposite directions in the presence of an electric field. I can summarize the remarks in the preceding paragraph by saying that elements on the left of the Periodic Table are likely to lose their few outermost electrons and so become cations; those on the right of the table with nearly completed outermost clouds are likely to gain electrons and so become anions.

We have come across one of the great bonding mechanisms: because opposite charges attract one another, and cations and anions are oppositely charged, it follows that atoms that form these ions will clump together into a compound. Common salt, sodium chloride, is an excellent example of this type of compound formation. Sodium (Na, from its Latin name *natrium*) lies on the left of the table, and readily releases its single outermost electron to form a sodium cation, denoted Na^+. Chlorine (Cl) lies on the right of the table, and happily accommodates an additional electron to complete its outer layer and thereby become a chloride anion, Cl^-. (Note the tiny change of name from chlorine to chloride). The ions clump together, and form sodium chloride, a solid rigid mass of ions held together by their mutual attraction. I have already emphasized that atoms are very small, and that even tiny samples of a substance contain a lot of them. You are an Atlas among stars when it comes to ions, for when you pick up a grain of salt, you are holding more ions than there are stars in the visible universe.

You are now in a position to see why salt mined in one place or extracted from a sea somewhere has the same composition as another sample mined or obtained on the other side of the world. A sodium atom has one electron in its outer layer; a chlorine atom has a single vacancy in its; so the only combination possible is for one sodium atom to bond with one chlorine atom by this process of giving up and acquiring electrons to become ions. Universally, common salt is NaCl with sodium and chloride ions present in the ratio 1:1. Compounds like Na_2Cl (ions present in the ratio 2:1) or

Na_2Cl_3 (ions present in the ratio 2:3) and so on simply can't exist. It should be becoming apparent that Nature has rules about which liaisons can form and which cannot.

The type of bonding that I have described so far is called *ionic bonding*. It typically results in rigid, brittle solids that melt only at high temperatures. The granite and limestone of our landscapes are examples of materials composed of atoms held together by ionic bonds. That we do not sink through either when we stand on them can be traced to the fact that the layers of electrons round the nuclei of their atoms, now present as ions, are complete, and the clouds of our atoms cannot occupy the same space as the clouds of their atoms. Our bones are also largely ionic, and provide a reasonably rigid framework for our organs.

Our squishy organs, our flesh, the coating of our flesh in fabrics, the fabric-analogue coating of limestone by vegetation, the upholstering of our landscapes, are all clearly of a different character. Although ions might be present, they are not responsible for the major character of these structures. Here we are in the realm where atoms are held together by completing their cloud layers by *sharing* electrons. This type of bonding is called *covalent bonding*, the 'co' indicating cooperation and the 'valent' derived from the Latin word for strength: *Valete*! was the Roman 'Goodbye! Be strong!'.

A simple example of covalent bonding is that responsible for the structure of a water molecule, which just about everyone knows is H_2O. Oxygen, with its six outermost electrons can accommodate two more electrons to complete its outermost cloud layer (which can hold, remember, a maximum of eight electrons). A hydrogen atom can provide one electron, and can complete its own outermost cloud layer (the only one it has) by acquiring one more electron (that first, innermost layer, remember, can accommodate only two electrons). Sharing can be complete provided two hydrogen atoms are content to share two electrons with oxygen: the hydrogen atoms each get a share in two electrons and the oxygen atom gets a share in eight electrons. At once, we see that water cannot be H_3O or HO_2: H_2O is the only bonding pattern that results in complete outermost cloud layers for all the atoms. Ammonia, NH_3 (where N denotes nitrogen) also falls into place, because a nitrogen atom has five electrons in its outer layer, and so needs three more to complete its layer. That is satisfied by the presence of three hydrogen atoms willing to share an electron each. Methane, CH_4, falls into place too, because carbon has four vacancies.

You, like chemists, need to be aware of one very important distinction between ionic and covalent bonding. Ionic bonding results in huge aggregates of ions: essentially chunks of substance. Covalent bonding commonly results in discrete atomic assemblies, like H_2O. That is, covalent bonding results in individual molecules. This distinction is hugely important, and you need to keep it in mind. It is for this reason that all gases are molecular, such as oxygen (O_2 molecules) and carbon dioxide (CO_2 molecules); there is no such thing as an ionically bonded gas! Even if such a gas were formed, all the ions would immediately clump together as a solid. Just about all substances that are liquid at normal temperatures are molecular, as the molecules need to be able to move past one another and not be trapped in place by a strong attraction to their neighbours. Water is an obvious example; gasoline another.

Covalent bonding can result in solids, so you should not infer that every solid is ionic: all ionic compounds are solids at ordinary temperatures but not all solids are ionic. An example of a covalently bonded solid compound is sucrose, a covalent compound of carbon, oxygen, and hydrogen with the composition $C_{12}H_{22}O_{11}$ with the atoms in each molecule linked together by covalent bonds into an intricate web.

One very important aspect of covalent bond formation is the overriding importance of pairs of electrons. One of the greatest chemists of the 20th century, Gilbert Lewis (1875–1946), identified its importance, but it remained for quantum mechanics to provide an explanation. As far as we are concerned, each shared pair of electrons counts as one covalent bond, so it is easy to count the number of bonds that any atom has formed simply by counting the number of pairs of electrons that they share. One shared pair counts as a 'single bond' (denoted –), two shared pairs between the same two atoms counts as a 'double bond' (denoted =), and three shared pairs counts as a 'triple bond' (denoted ≡). Only very rarely does sharing proceed any further, and so these three types of sharing are all we need to know about. Each hydrogen atom in H_2O is joined to the oxygen atom by a single bond. Carbon dioxide is a molecule with two sets of double bonds, and can be denoted O=C=O. Triple bonds are much rarer, and I shall not discuss them further except to mention that the gas acetylene of oxyacetylene welding, H–C≡C–H, is an example.

The question lurking behind this account is why two electrons (an 'electron pair') are so fundamental to covalent bond formation. The explanation lies deep in quantum mechanics. A hint of the reason is that all electrons spin on their axis. If two electrons lock their spins together by rotating in opposite directions, then they can achieve a lower energy. Another manifestation of the importance of this spin-locking is the fact that, as we have seen, the cloud layers each hold an even number of electrons (2, 8, etc.). The French words for an unpaired electron, an *electron célibataire*, is a perhaps typically Gallic allusion to the importance of pairing.

Metals

I have concealed from view so far the existence of a third type of bond. The majority of elements are metals: think iron, aluminium, copper, silver, and gold, and metals play a very special role in chemistry, as we shall see. A block of metal consists of a slab of atoms, but are those atoms held together by ionic or covalent bonds? We are immediately confronted by a problem. All the atoms in the block are the same, so it is unlikely that half will form cations and the other half anions, so ionic bonding is ruled out. If all the atoms were bonded covalently, we would expect a rigid solid (like diamond, in which the carbon atoms are in fact so bound); but metals can be beaten into different shapes (they are 'malleable') and drawn out into wires (they are 'ductile'). They are also lustrous (reflective of light) and conduct electric currents, a stream of electrons.

Metal atoms are bound together by *metallic bonding*. That is not just a tautology. The clue to its nature is the fact that all the metals lie towards the left-hand side of the Periodic Table where, as we have seen, the atoms of the elements have only a few electrons in their outermost cloud layers and which are readily lost. To envisage metallic bonding, think of all these outermost electrons as slipping off the parent atom and congregating in a sea that pervades the whole slab of atoms. The cations that are left behind lie in this sea and interact favourably with it. As a result, all the cations are bound together in a solid mass. That mass is malleable because, like an actual sea, it can respond readily to a shift in the positions of the cations in the mass when they are struck by a hammer. The electrons also allow the metal to be drawn out into a wire, by responding immediately to the relocation of the cations. As the electrons in the sea are not pinned down to particular atoms, they are mobile and can migrate through the solid in response to an electric field. Metals are lustrous because the electrons of the sea can respond to the shaking caused by the electric field of an incident ray of light, and that oscillation of the sea in turn generates light that we perceive as reflection. When we gaze into the metal coating of a mirror, we are watching the waves in the metal's electron sea.

The chemistry lesson at this stage in our account is that the elements that are metals in their natural state are the ones that can readily lose electrons from their outer layers. These elements are therefore also the elements that form cations when anion-formers are present and able to accept the discarded electrons. Elements on the far right of the Periodic Table are electron acquirers, as they have one or two gaps in their outer cloud layers and can accommodate incoming electrons, those donated by the atoms that form cations. Ionic compounds (bear in mind sodium chloride) are therefore typically formed between a metallic element on the left of the table with a non-metallic element on the right of the table.

With that summary in mind, you are starting to think like a chemist, being able to anticipate the type of compound that a combination of elements is likely to be, and beginning to anticipate its properties. You are also beginning to understand how the Periodic Table relates to the properties of the elements and the compounds they form, and how the family relationships between neighbours, which spring from the cloudlike electrons and the periodic repetition of analogous arrangements, are displayed in practice.

Where we are, and the next step

Such are the central principles of chemistry as far as structures are concerned. They boil down to the existence of atoms, an acknowledgement of their structures, and the behaviour of electrons. Our next concern is with the 'carrot and the cart' of chemistry: energy.

Chapter 3
Its principles: energy and entropy

Atoms are one great river of understanding in chemistry; the other river consists of energy. To understand why and how reactions take place and why and how bonds form in all their variety, chemists think about the energy changes that take place when processes occur. Chemists are also interested in energy for its own sake, as when a fuel is burned or food, a biological fuel, is deployed in an organism. As I remarked in Chapter 1, the study of energy and the changes that it can undergo is the world of thermodynamics, to which we now turn.

I have written extensively on the laws of thermodynamics and do not intend to reprise my discussion here. As I did for quantum mechanics in Chapter 2, I shall distil the essence of what is necessary and which chemists typically keep in mind or at least the back of their minds while going about their business.

Some thermodynamics

The essence of chemical thermodynamics is that there are two aspects of energy that it is necessary to keep in mind: its quantity and its quality. The *First Law* of thermodynamics asserts that the total energy of the universe is constant and cannot be changed. The energy can be parcelled out in different ways and converted from one form to another, but no process can change its total quantity. Thus, the First Law sets the legal boundaries for change: no change can occur that would alter the total amount of energy in the universe. The *Second Law* of thermodynamics asserts that the quality of energy degrades in any natural change. This law is expressed more formally in terms of the *entropy*, a measure of the quality of energy in the sense that the higher the entropy the lower is its quality, and stated as 'the entropy of the universe tends to increase'. In a refined meaning of 'disorder', entropy is a measure of disorder, with greater disorder implying greater entropy. The Second Law can be regarded as a summary of the driving power of natural change, including chemical reactions, for only reactions that result in the degradation of the quality of the total energy of the universe can occur naturally. In short, with increasing disorder in mind, things get worse. A summary of thermodynamics, the core of its essence, is therefore that the First Law identifies the feasible changes from among all possible changes (no change in total energy) and the Second Law identifies the natural changes from among those feasible changes (the entropy must increase).

The role of energy

Chemists deploy these two concepts in a variety of ways. In their conventional thinking they adopt the view that bonds form or are replaced by new bonds in the course of a reaction if that reorganization of atoms results in a reduction in energy. That remark, though, to a fusspot like me, is quite wrong, but like many false statements it is a handy and memorable rule-of-thumb. It is wrong because the legal authority of the First Law rules against it: the total energy cannot change. The correct explanation is that if a process, such as bond formation, releases energy into the surroundings, then that represents a degradation of energy as it spreads and becomes less readily available: the release increases the entropy of the universe and so is a natural process. That the rule-of-thumb 'it lowers the energy' works most of the time is due to the fact that the spread of energy so released results in an increase in entropy. Working chemists quite sensibly use the rule-of-thumb all the time and I shall follow them. However, I shall keep my fingers crossed when I use it, and inwardly say to myself, a little like Galileo's whispered apocryphal *eppur si muove* ('and yet it moves') concerning the motion of the Earth around the Sun, that it is really entropy going up rather than energy going down.

A bond between atoms forms if (fingers crossed) it results in a reduction in energy. The type of bond that forms, ionic (attraction between ions) or covalent (shared electron pairs), depends on whether more energy is released by the total transfer of an electron from one atom to the other to result in ions, or by partial release and sharing. Thus, whether two elements form an ionic or a covalent compound can be assessed by considering the energy changes that accompany the various types of bond formation.

The same is true of the characteristic *valence* of an element, the typical number of covalent bonds that it can form. Valence is another aspect of its chemical personality and family relationship to its neighbours and is implied by its location in the Periodic Table. We saw in Chapter 2 that oxygen, with its two gaps in its outermost cloud layer, can complete that layer by reaching agreement with two hydrogen atoms to form H_2O, specifically H–O–H, indicating a valence of 2. Any further attachment of hydrogen atoms would require electrons to occupy a new outer cloud layer far from the nucleus, and there would be no energy advantage in doing so. Forming fewer bonds would not reap the advantage of forming two. Therefore, on energetic grounds, the valence of oxygen is expected to be 2. Also in Chapter 2 we saw another example of that valence in oxygen's combination with carbon in carbon dioxide, CO_2, specifically O=C=O, where it also displays a valence of 2. As can be seen in this case too, carbon displays its typical valence of 4, just as it does in methane, CH_4.

Now we can see how the location of an element in the Periodic Table indicates its characteristic valence: carbon's typical valence is 4, its neighbour nitrogen is 3, and nitrogen's neighbour oxygen is 2. Much the same can be said of their neighbours in the row below: silicon's valence is typically 4, phosphorus's is 3, and sulfur's is 2. Once again, we are seeing how energy considerations in collaboration with concepts of atomic structure—particularly the completion of their cloud like layer structure—accounts for similarities between neighbouring elements.

Keeping track of energy

Energy is released in many chemical reactions, as in the combustion of natural gas or gasoline. The process is not simply the release of energy when bonds form, because the starting materials, such as methane, already have atoms bonded together. In many reactions, and here I shall focus on combustion, bonds must be broken and new bonds formed. The energy released is the difference of the two contributions. For instance, in the combustion of methane, due to its reaction with oxygen, O_2, the four carbon–hydrogen bonds of methane and the bonds linking the two oxygen atoms in oxygen must all be ripped apart, which takes a lot of energy, before new carbon–oxygen bonds in carbon dioxide and hydrogen–oxygen bonds in water are formed, which releases energy. Only if the energy released in the subsequent bond formation exceeds the energy required for initial bond-breaking will the combustion release energy as heat. If the balance were the other way round, burning methane would result in refrigeration!

Chemists use thermodynamics to keep track of these individual changes in energy, and to assess the net change that takes place in a reaction. For this purpose, they use an assessment of the quantity of energy available from a reaction as heat that is called the *enthalpy*. The name comes, evocatively, from the Greek words for 'heat inside'. There are good technical grounds for distinguishing enthalpy from energy, but for our purposes we can think of enthalpy as just another name for the energy trapped in compounds and available as heat.

In a so-called *exothermic reaction*, energy is released as heat and the store of enthalpy decreases. All combustions are exothermic, and in the combustion of methane the enthalpy of methane + oxygen falls to the enthalpy of carbon dioxide + water, the difference escaping as heat. Chemists assess the efficiency of fuels by considering the enthalpy changes that accompany their combustion, with full reservoirs of enthalpy being preferred as more heat is available from a given amount of fuel. The study of enthalpy and the release of heat in chemical reactions is called *thermochemistry*. It makes a substantial contribution to our understanding of foods and fuels and is also used to gather data for more general thermodynamic discussions.

Most reactions, not only combustions, are exothermic, with the starting materials collapsing into the lower-enthalpy products of the reaction and thereby achieving lower enthalpy overall. It is perhaps easy to understand that many reactions proceed in the direction of lower enthalpy, just as the finger-crossing rule-of-thumb about energy suggests. However, here is a puzzle, a puzzle that left 19th century chemists totally nonplussed: some reactions move upwards in enthalpy naturally. Reactions that absorb heat and increase their store of enthalpy are called *endothermic reactions*. There are not many common ones that occur naturally, but the fact that there is even one raised the collective puzzled eyebrows of the 19th century chemists, for how, they wondered, can anything run naturally uphill, in this case, uphill in enthalpy?

They didn't know about entropy, and they took literally the rule-of-thumb about things falling naturally to lower energy. Chemists now know that entropy determines the direction of reactions, and *provided the entropy increases*, the reaction can either travel uphill or downhill in enthalpy. To understand why, we have to remember that entropy is a measure of the quality of energy.

When energy spreads into the surroundings of a reaction flask and becomes dispersed, the entropy goes up, so it should be easy to understand why exothermic reactions are so common. However, we need to think about what is going on inside the flask. Suppose that in the course of a reaction energy flows into the flask: the entropy goes down because now the energy is localized, less dispersed, and more readily available: it has become of higher quality. Suppose, though, that at the same time a great deal of disorder is generated within the flask. Now the total entropy of the universe might increase despite the energy becoming more localized. If that happens, then the endothermic reaction will occur naturally.

Where the 19th century chemists went wrong was to suppose that, like Newton's apple, reactions rolled down in enthalpy; what 21st century chemists know is that reactions roll up in entropy: disorder increases; things get worse. Often the two lead to the same conclusion, but in all cases entropy is the property to consider. Increasing entropy is the signpost of change, and sometimes it points in an endothermic direction. If you still continue to want to think, from your familiarity with gravity, that the natural direction of change is 'down', then think that natural change it is invariably down in the quality of energy.

The rates of reactions

We now know where a reaction goes: the signpost of natural change is towards higher entropy of the universe, the degradation of energy. There are two associated questions. One is how fast it goes to wherever it is going, and the second is what route it takes to get there. I shall deal with the first question here and tackle the second in Chapter 4.

Chemists take a great deal of interest in the rates of chemical reactions as there is little point in knowing that they can, in principle, generate a substance in a reaction but that it would take them millennia to make a milligram. The study of reaction rates is called *chemical kinetics*. We shall see that energy is a crucial component of the explanation of the wide range of rates that are observed. That range is indeed very wide: some reactions are complete in fractions of seconds (think explosions); others take years (think corrosion).

Chemists measure reaction rates in a straightforward way, by monitoring the change in amount of a product over time. They do these measurements for a variety of reasons. One, the most basic, is simply to know what concentration to expect at any given moment. More significantly, especially for industrial applications, they may wish to find the conditions that result in products being formed at the optimum rate. A third reason is to discover what is called the *mechanism* of the reaction, the sequence of changes at an atomic level that converts the starting materials, the 'reactants', into the final product. Very detailed information of the last kind is obtained by firing one stream of molecules at another and monitoring the outcome of the collisions that take place.

My concern here is with the role of energy in determining the rate of a reaction. We have seen that there may be a natural tendency for a reaction to occur, so the question arises why all reactions aren't over in a flash. This question is supremely important, for the slow, restrained development of products in many cases allows for the subtle operations that constitute life: if biological reactions were all over in a flash we would all instantly be goo.

Chemists have identified the existence of a barrier to instant reaction. By making measurements on the effect of temperature on the rates of reactions, they have identified the need for molecules to acquire at least a minimum energy, called the *activation energy*, before the atoms of the reactants are able to rearrange into products. This requirement is easiest to understand for reactions in gases, where molecules are ceaselessly undergoing collisions with one another with various energies of impact. Only highly energetic impacts between really fast molecules bring enough energy to loosen the bonds holding atoms in their initial arrangements and enabling them to settle into new ones. As the temperature is raised, molecules move faster and a higher proportion of the collisions occur with at least this minimum energy, so the rate of the reaction increases. Some activation barriers are very high, and hardly any collisions are sufficiently energetic to result in reaction at normal temperatures. The reaction of hydrogen and oxygen is an example: the two gases can be stored together indefinitely at normal temperatures but explode at high temperatures or when a spark provides sufficient energy locally to set the reaction in train.

Much the same requirement of a minimum energy applies to reactions in solution too, including those in the watery interiors of living things. In this environment, molecules do not hurtle through space and collide: they jostle through the fluid, meet, and might jostle away unchanged. However, there is a chance that when two reactant molecules are together, they are jostled so violently by the surrounding water molecules that their atoms are eased apart and can rearrange into products. The chance that sufficiently violent jostling will occur increases sharply with increasing temperature, so even reactions in fluid environments go faster when they are heated. Fireflies, for instance, flash more rapidly on warm nights than on cool nights; we heat to induce the reactions in the kitchen that we call 'cooking' foods.

In many instances a reaction can be made to go faster by introducing a *catalyst*, a substance that increases the rate of the reaction but is otherwise unchanged. The Chinese characters for catalyst form the word 'marriage broker', which captures the sense of its role very well. A catalyst acts by providing a different pathway—a different sequence of atom migrations and bond formations—for a reaction, a pathway with a lower activation barrier. Because the activation energy is lower, more successful encounters between reactants take place at ordinary temperatures and the reaction is faster. Catalysts are the lifeblood of the chemical industry, where the efficient, rapid production of desired substances is essential and the success of an entire industry depends on the identification of the appropriate catalyst. A point to note is that there is no such thing as a 'universal catalyst', and each reaction must be studied individually and an appropriate catalyst devised. Another point is that not all reactions can be catalysed: in many cases we have to live with Nature's decision about the rate.

Catalysts are essential for the functioning of our bodies. *Enzymes* (a word derived from the Greek word *zyme*, to leaven) are protein molecules that function as catalysts and control with considerable specificity and effectiveness just about all the chemical reactions going on inside us. Life is the embodiment of catalysis.

The nature of equilibrium

A very important aspect of reaction rates concerns what is going on when a reaction has completed and change is no longer apparent. Chemists say that the reaction has reached *equilibrium*. In many cases, barely a single molecule of the starting material remains, but in many cases the reaction seems to stop before the starting materials have all been used. An example of the latter is the hugely economically significant reaction between nitrogen and hydrogen for the synthesis of ammonia (NH_3) in the 'Haber–Bosch process', which lies at the head of processes that include the manufacture of much of the world's agricultural fertilizer. That reaction seems to come to a stubborn stop with only a small fraction of the nitrogen and hydrogen converted into ammonia, and however long we wait, and however much catalyst we shovel in, no further change occurs. The reaction has reached equilibrium.

Equilibrium is only an apparent cessation of reaction. If we could monitor an equilibrium mixture at an atomic level, we would find that it is still a turmoil of chemical activity. Products are still being formed when a reaction is at equilibrium, but they are decaying back into the starting material at a matching rate. That is, chemical equilibrium is a *dynamic* equilibrium, in which forward and reverse processes are occurring at matching rates so that there is no net change. In the synthesis of ammonia, ammonia molecules are still being formed at equilibrium, but are being ripped apart into nitrogen and hydrogen at the same rate as they are being formed and there is no net change.

The important consequence of chemical equilibrium being dynamic and not just dead is that it remains responsive to changes in the conditions. Thus, even though certain reactions inside our bodies might have reached equilibrium, they are responsive to changes in temperature and other factors, and it is that responsiveness that keeps us alive. 'Homeostasis', the delicate and complex balance that keeps bodies alive and alert, is a manifestation of this dynamic, responsive, chemical equilibrium. As to the industrially all-important synthesis of ammonia, the fact that the equilibrium is dynamic rather than dead gives chemists and industry hope that perhaps the equilibrium can be manipulated and the yield of ammonia improved. That was the prospect confronting the chemist Fritz Haber (1868–1934) and the chemical engineer Carl Bosch (1874–1940) back at the beginning of the 20th century, who in due course discovered that with an adroit choice of catalyst and by working at high pressures and temperatures, they could bend the equilibrium to their will. In so doing, they fed the world.

Where we are, and where we are going

We have now seen that energy is both the carrot and the cart of chemical reactions, and so can finally unwrap the meaning of my delphic remark at the end of Chapter 2. Energy, its dispersal in disorder, is the carrot: the driving power of chemical reactions. Energy, the need to overcome the barriers between reactants and products, is also the cart, in the sense of holding back free unrestrained flight towards the carrot.

I have said hardly anything about how reactants actually undergo the atomic rearrangement that leads to products. Unravelling and understanding those changes, and making use of them to bring about amazing and almost magical transformations, lies at the heart of practical chemistry, and is the next step in our journey.

Chapter 4
Its reactions

Whenever anyone thinks of chemistry, they think of its reactions, reactions that flash, bang, change colour, or stink. They are aware that reactions go on in chemical plants, that the combustion of a fuel and the manufacture of plastic, paint, or pharmaceuticals are reactions. Perhaps some correctly think of cooking as causing reactions, and most are probably at least vaguely aware that we ourselves are elaborate test-tubes who are alive as a result of the myriad reactions within us. But exactly what are reactions? What is going on when chemists stir and boil their liquid mixtures, pour one liquid into another, and generally go about their seemingly arcane activities in laboratories?

They are coaxing atoms to exchange partners. The starting stuff, the 'reactants', consists of atoms in one state of combination; the stuff that is produced, the 'products', consists of the same atoms but in a different state of combination. The shaking, stirring, and boiling is bringing about that change from one state of combination to another, prising atoms apart in one kind of molecule and encouraging them to form different kinds of molecules. In some cases, the atoms of the reactants immediately tumble into the desired new arrangement, whereas in others, the chemist must scheme and seduce, devising elaborate coaxings through a sequence of subtle steps. Combustion and explosion might stem from a spark; to generate an intricate weblike pharmaceutical molecule might take thought, luck, time, and careful, sophisticated, erudite planning.

A chemical laboratory is full of specialized equipment, a lot of which is there to determine whether the product of a reaction is what the chemist hopes or thinks it is. I explain some of its functions in Chapter 5. A lot of it is directly involved in the business of atom-coaxing and separating the chemical wheat from the chaff, the desired product from the waste. There are test-tubes, flasks, beakers, distillation apparatus, filtering apparatus, and various heaters, shakers, and stirrers. Despite this bewildering (and expensive) array of apparatus, through a chemist's eyes there are only a small number of processes going on at an atomic level: four, to be precise. It is in fact worth pausing at that remark, and to realize that all the wonders of the world, both natural and synthetic, are spun from a handful of elements and four ways of manipulating them.

In the rest of this chapter I shall introduce you to those four fundamental types of reaction. In some cases, they conspire together and their collaboration at first sight seems to be of a new type of reaction, but when that conspiracy is picked apart, there they are.

Proton transfer: acids and bases

Chemists discovered the fundamental particle known as the 'proton' long before the physicists had pinned it down, but the chemists did not realize that they had done so. A proton, remember from Chapter 2, is the tiny, singly charged nucleus of a hydrogen atom. Its low charge (which means that it is often only loosely gripped by a neighbouring atom in a molecule) and low mass (which makes it nimble) mean that a hydrogen atom that is part of a molecule might suddenly find that its nucleus, the proton, has slipped away and become embedded in the electron clouds of a more welcoming nearby molecule. That—the transfer of a proton from one molecule to another—in a nutshell, is one of the four great fundamental types of reaction.

We are in the world of acids and alkalis. Although the early chemists were familiar with acids, it took them a long time to realize that an acid is a compound with hydrogen atoms that have little control over their nuclei and are apt to lose them. Acids at one time were recognized, as their name suggests (Latin *acidus*: sour, sharp), by their sharp taste. Chemists who survived that hazardous test (now existing in a more palatable way in our response to the tang of vinegar, soda, and cola drinks) had no idea that what was tickling their tastebuds were protons. That recognition came as late as 1923, when the British chemist Thomas Lowry (1874–1936) and the Danish chemist Johannes Brønsted (1879–1947) independently proposed that an acid is any molecule or ion that contains hydrogen atoms that can release their proton nucleus to another molecule or ion. Not all molecules that contain hydrogen can act in this way, as the proton may be too heavily embedded in the electron clouds, but various classes of molecule can, especially if other atoms in the molecule can draw the electron cloud away from the proton and enable it to escape. Acetic acid, the acid in vinegar, is one such compound; others include hydrochloric acid (HCl) and sulfuric acid (H_2SO_4). If you ever see H written first in a formula, that is an indication that it can release its proton and act as an acid. (What about H_2O?, you might be thinking: wait and see).

One hand cannot clap alone. If there is a proton donor (an acid), presumably there must be a proton acceptor, a molecule or ion to which the liberated proton can attach and burrow itself into the electron cloud. This is where alkalis come in (the name comes from the Arabic *al qaliy*, the ashes, for wood ash is a source of alkali).

The test for an alkali used to be just as hazardous as that for an acid: in this case, an alkali has a soapy feel. We now know that alkalis turn fats into soaps, so in the test the fats on the finger of the tester were being turned into soap. Needless to say, chemists have more survivable and sophisticated tests now. The underlying reason for the ability of alkalis to turn fats into soap is the presence in them of *hydroxide ions*, OH^-, which are species that can attract and keep protons, in the course of which becoming water molecules, H_2O.

Here is a tiny technical point that I really do need to introduce. Chemists now refer to a proton-accepting molecule and ion as a 'base'. Thus, OH^- is a base. They keep the term 'alkali' for bases dissolved in water. So, for instance, sodium hydroxide, $NaOH$, dissolves in water, separating into Na^+ ions and OH^- ions. It is therefore a source of the base OH^- and the solution is an alkali. I shall use the term 'base' from now on, because it is more general than alkali (a molecule or ion doesn't need to be present in water to be a base).

Why the name 'base'? When hydrochloric acid reacts with sodium hydroxide solution, salt (sodium chloride) and water are formed when the acid's proton skips across on to the OH^- ion provided by the sodium hydroxide. When instead sulfuric acid reacts with sodium hydroxide solution, sodium sulfate and water are formed when the acid's proton skips across on to the OH^- ion provided by the sodium hydroxide. We are building different compounds, sodium chloride and sodium sulfate, on foundations of the same base, sodium hydroxide: hence the name.

Incidentally, both sodium chloride and sodium sulfate are called *salts*, the general class of these ionic substances formed by the reaction of an acid and a base taking its name from a common exemplar, namely common salt, sodium chloride. That is a common feature in chemistry, where the name of one type of compound inspires the name of a whole related class.

A large number of reactions are reactions between acids and bases, their common feature being that a proton is transferred from the acid to the base. Among the most important are reactions going on inside organisms, including corn, oak trees, flies, frogs, and us, for many enzyme-controlled biochemical reactions, such as those involved in the metabolism of food and respiration, are of this kind. In fact, you could regard life as one long, highly elaborate titration!

One reason for the importance of proton transfer, acid–base reactions is that the presence of the arriving proton with its positive charge distorts the electron cloud of the base, perhaps exposing an atom nearby in the molecule to attack by other atoms as the electron cloud around it is pulled away. Thus, proton transfer prepares atoms and the bonds that hold them to attack and then further reaction. This preparation for attack is a major role of acid–base reactions in our bodies, with enzymes preparing smaller molecules for digestion or modification.

I suspected earlier that you might worry about water and its formula H_2O, which with its leading H atoms might suggest that it is an acid. It is. When you drink water you are drinking almost 100 per cent acid. Water is also a base. You should know that when drinking water you are drinking almost 100 per cent pure base. I need to explain this alarming revelation. Although alarming, the acceptance of the fact that water is both an acid and a base is central to the way that chemists think about it, the solutions it forms, and the reactions it undergoes.

Think of yourself as a water molecule in a glass of water, surrounded by other water molecules in a dense, jostling crowd. One of your hydrogen atom's protons can slip out of you and stick on to a neighbour. That transfer implies that, being a proton donor, you are an acid. Your neighbour, who accepted the proton, is behaving as a base. The loss of a proton leaves you as an OH^- ion, a hydroxide ion; the gain of a proton makes your neighbour an H_3O^+ ion, which is called a *hydronium ion*. The incoming proton is like a hot potato, and it is immediately passed on to one of your neighbour's neighbours. Likewise, your negative charge can pull a proton out of one of your neighbours, rendering you H_2O again. This ceaseless turmoil of passing on a proton and flickering from OH^- to H_2O to H_3O^+ goes on throughout the liquid. The actual concentrations of the OH^- and H_3O^+ ions is very, very small, so when you look at a glass of water you should think of it as overwhelmingly H_2O molecules, but with just a few OH^- and H_3O^+ ions throughout it, with identities that are ceaselessly changing as protons hop around. Each H_2O molecule, though, is an acid (a proton donor), and each H_2O molecule is a base; that is why I said that water is a nearly pure acid and a nearly pure base.

Electron transfer: oxidation and reduction

The electron was discovered by the physicist J. J. Thomson in 1897. Chemists had unwittingly shifted it around for decades before that, with Michael Faraday (1791–1867) its arch-shifter: clearly, even he did not know what he was doing. The transfer of an electron from one molecule to another is the second of the four great fundamental reactions, with a great deal stemming from the migration of this little fundamental particle. Electron transfer, for instance, is the basis of great industries, such as steel-making. It is also responsible for the collapse of their artefacts through corrosion.

I need to introduce you to the terms 'oxidation' and 'reduction', and then explain how electron transfer plays a role in them. Oxidation sounds as though it means what it says: namely reaction with oxygen. However, although in science a term might have started life in common usage, it often captures more of the landscape by becoming generalized. We saw that a moment ago in the generalization of the term 'salt' from a single exemplar to a whole class of related compounds. So it is with oxidation too.

Let's take a simple example. Most of us have seen the bright light emitted when a strip of magnesium (Mg) burns in air. In this reaction, the magnesium metal combines with oxygen to form magnesium oxide, an ionic solid consisting of Mg^{2+} ions and O^{2-} ions. The energy released in the reaction is emitted as light and heat. The crucial point to note, however, is that each Mg atom of the metal has lost two electrons and has become a doubly charged Mg^{2+} ion. A similar reaction, but one far less familiar, occurs when magnesium burns in chlorine gas, when the product is magnesium chloride. That compound consists of Mg^{2+} ions and Cl^- ions. As in the first reaction, the crucial change is that each Mg atom has lost two electrons to become an Mg^{2+} ion. No oxygen is involved in the second reaction, but the same process, the removal of electrons, has occurred. Chemists now regard the second reaction simply as an oxidation of a general kind, and define oxidation as *the loss of electrons*. It is sometimes quite difficult to identify electron loss, such as in the combustion of a hydrocarbon fuel, but they have ways of doing so, and whenever electron loss takes place they call it an oxidation even though oxygen itself might not be involved in any way.

We saw when discussing the reactions between acids and bases that one hand cannot clap alone: if there is a proton donor (the acid), there must be a proton acceptor (the base). One hand cannot clap alone in an electron transfer reaction either, and the electrons lost in an oxidation must end up somewhere. That is where reduction comes in.

In the old days (I am being deliberately vague), reduction referred to the extraction of a metal from its ore: the ore was *reduced* to the metal. This process occurred, for instance, in that great hulking giant icon of the industrial revolution, the blast furnace, in which iron ore (an oxide of iron) reacts with carbon and carbon monoxide to form the molten iron (Fe, from the Latin *ferrum*) that dribbled out of the base of the furnace and went on for a future as various kinds of steel. Iron oxide consists of Fe^{3+} ions and O^{2-} ions. Iron metal consists of Fe atoms. With that in mind it is easy to see what has happened in the reduction of the ore:

electrons have attached to each Fe^{3+} ion to neutralize its charge and form Fe atoms.

The attachment of electrons to an atom is now taken as the definition of reduction, even though the reaction might have nothing (except that feature) to do with the reduction of an ore to a metal. Thus, in the combustion of magnesium in oxygen, the oxygen molecules receive the electrons released in the oxidation of magnesium and become O^{2-} ions: the oxygen is reduced. In the oxidation of magnesium by chlorine, the chlorine molecules receive the released electrons and become Cl^- ions: the chlorine is reduced. Whenever electrons released are transferred to an atom, that atom is said to be reduced.

We now have both hands clapping in an electron transfer reaction: oxidation (electron loss) always occurs with reduction (electron gain). Chemists recognize the need for these two hands and commonly refer not to an oxidation reaction alone, nor to a reduction reaction alone, but to a 'redox' reaction. (They don't, so far, extend that type of compressed naming to 'basid' reactions, combined base and acid hand-clapping).

Redox reactions are hugely important. We have already seen that they stand at the head of the steel chain, when iron is won from its ores. The reverse of that winning is the process of corrosion, when the ion artefacts are lost in the redox reactions that we call corrosion: when iron is oxidized by water and the oxygen of the air and reverts to its oxide. The combustion reactions that drive our vehicles are redox reactions, in which the hydrocarbon fuel is oxidized to carbon dioxide and water by reaction with oxygen (which is itself reduced).

The reactions that take place in the batteries that power our laptops, tablets, phones, and increasingly vehicles are redox reactions. Batteries are so important for the modern world as portable sources of electric current that it is worth a moment or two to understand the general principle of their operation and how they harness redox reactions.

We have seen that in an oxidation electrons are released and that in a reduction they are acquired. In a battery, the release and acquisition are spatially separated. Electrons are released into an electrode, a metallic contact, in one region of the battery, travel through an external circuit, and then attach to the species undergoing reduction at a second electrode elsewhere in the battery. Thus, the redox reaction, the joint oxidation and reduction reactions, proceed, and in doing so, the flow of electrons from one electrode to the other is used to drive whatever electrical equipment is attached to the device. Modern batteries use a range of redox reactions to bring about this electron flow, ranging from the heavy lead–acid batteries in vehicles to the light lithium-ion batteries in laptops, tablets, and phones.

Redox reactions can also be forced to take place against their natural direction by driving electrons into a reaction mixture through an electrode. This is the process of *electrolysis*, the process of causing chemical reaction by passing an electric current. Electrolysis is the principal way of extracting aluminium (Al) from aluminium oxide. A powerful current is forced into a cell containing aluminium oxide dissolved in a special solvent, and the electrons that enter the cell are forced on to the Al^{3+} ions, forming Al atoms. Electrolysis is also used to purify copper and to deposit metals, such as chromium, on to the surfaces of other metals.

One feature that distinguishes the electron transfer of redox reactions from the proton transfer of acid–base reactions is that because electrons are intimately involved in bonding, the migration of an electron from one molecule to another can drag with it several other atoms. We have seen a little of that, without drawing attention to it, in combustion reactions, where in the course of the oxidation of a hydrocarbon molecule, carbon, oxygen, and hydrogen atoms are dragged around as the electrons migrate between the molecules, and the hydrocarbon and oxygen molecules are reassembled into carbon dioxide and water molecules. This difference is hugely important in the reactions of organic chemistry, where clever use of atom-dragging redox reactions can be used to construct intricate structures.

It is partly due to this ability of migrating electrons to drag atom baggage with them that redox reactions are so important in biology: they keep the biosphere (including that small part, you) alive and vibrant. Photosynthesis, the process by which sunlight is captured and used to power the formation of carbohydrates in green plants, is a chain of electron transfer reactions, which have the overall effect, when atom-dragging is taken into account, of using the hydrogen atoms of water and the carbon and oxygen atoms of carbon dioxide, to construct carbohydrates, including starch and cellulose. In the form of organic corrosion we call digestion, those redox-formed carbohydrates are mined for their carbon and hydrogen atoms in a sequence of redox reactions we call respiration and metabolism.

Radical reactions

The third kind of reaction takes place when radicals meet. You need to know that a *radical* (or 'free radical') is a molecule with an odd number of electrons. We have seen that electrons pair together when bonds form, so a radical is a molecule in which all the electrons except one have paired and hold the atoms together, and there remains one unpaired electron. A radical is commonly denoted R· or ·R, the dot representing the unpaired electron.

Most radicals are aggressively reactive and do not survive for long in the wild. In some cases, two radicals might collide, and clump together as their unpaired electrons pair and bind the two radicals together to form a conventional molecule with an even number of electrons: R· + ·R → R–R. This kind of process occurs in flames, which are environments rich in radicals because the stress of high temperatures rips molecules apart, sundering the electron-pair bonds. Indeed, one form of fire-retardant is a substance that gives rise to radicals when heated. These radicals, lets denote them ·X, latch on to the radicals that are propagating the flame and quench their chemical aggressiveness, R· + ·X → R–X, so that the flame peters out.

Other radicals are of great commercial importance, for they are involved in the formation of many plastics. The general idea behind this process, which is called *polymerization*, is that when a radical R· attacks an ordinary molecule M it might attach to it. However, the outcome is again a molecule, now RM·, with an odd number of electrons, so it is also a radical. That radical can go on to attack another M molecule and attach to it. The result is still a radical, now RMM·. In other words, there can be a *chain reaction*, a chain of processes that propagates indefinitely, to give a long snaking RMM . . . M· radical, or until two such radicals collide, stick together by electron pairing, and so terminate the chain.

Those ubiquitous plastics polythene (polyethylene), polystyrene, and PVC (polyvinyl chloride) are made in this way. In the case of polythene the molecule M, which in this context is called a *monomer*, is ethylene, $H_2C=CH_2$. The outcome of the polymerization is a long chain, a *polymer*, of hundreds of $-CH_2CH_2-$ units. Chemists have found that by starting with different versions of ethylene, such as $H_2C=CHX$, where X can be a group of atoms, they can form polymers with a wide range of properties. Thus, when X is a benzene ring, the polymer is polystyrene and when X is a chlorine atom the polymer is PVC. To obtain the non-stick Teflon® all the hydrogen atoms are replaced by fluorine atoms; that is why its more formal name is polytetrafluoroethylene (PTFE).

Another variety of acids: Lewis acids

The fourth and final type of fundamental reaction might seem arcane at first sight, but it is a seriously important process. We have just seen that two radicals can form a bond with each other if each brings along one electron, which then pair. In this final type of reaction, one molecule provides *both* electrons of the bond that forms between them, the other partner in the reaction accommodating both electrons. We could represent this sort of reaction by A + :B → A–B, where the double dot on B represents the electron pair that is en route to be shared with A. Reactions of this kind are called *Lewis acid–base reactions* after the American chemist G. N. Lewis, who first identified them and was later killed by them. (He died after ingesting cyanide ions, CN^-, a poison that acts by this kind of reaction). They are called 'acid–base' reactions because they show marked similarities to the acid–base reactions I discussed earlier, in which a proton migrates from an acid to a base. Indeed, they can be regarded as yet another generalization of the concepts of acid and base, but I will not take you down the fascinating scenery of that route.

One role of Lewis acid–base reactions is to bring colour to the world. This remark gives me the opportunity to introduce you to *transition metal complexes*, which are often brightly coloured and which are formed in what I shall call a Lewis way. The haemoglobin of your blood is an example.

A transition metal is one of the elements in the skinny central part of the Periodic Table, and includes iron (Fe), chromium (Cr, this name anticipates the colour to come as *chroma* is the Greek word for colour), cobalt (Co), and nickel (Ni). The ions these elements form, such as Fe^{2+} and Co^{3+}, are commonly found surrounded by and bonded to six small molecules and ions that have an independent existence, such as H_2O, NH_3, and CN^-. These species are called *ligands* and the complete clusters are called *complexes*.

A complex is held together by bonds formed by the sharing of a pair of electrons provided by each ligand, so the metal ion is acting as the Lewis acid (the A) and each ligand is acting as the Lewis base (the :B).

In water, transition metal ions are typically surrounded by six water molecules acting as Lewis bases. When another Lewis base is added to the solution, it might drive out one or more water molecules and take their place. The electronic structure of the resulting complex might be quite different from that of the original complex, and as a result be brightly coloured. Many pigments and dyes are complexes formed in this way.

Breathing is a Lewis acid–base reaction. The oxygen carrier in our blood is haemoglobin, a huge protein molecule that has embedded in it four iron ions. Each one is gripped in place by four nitrogen atoms belonging to the protein framework and lying round it at the corners of a square. The bonds between the iron and the nitrogen atoms are the result of Lewis acid–base interactions, with Fe^{2+} the acid and each :N a base. When you breathe in, this already Lewis-constructed entity takes part in another Lewis acid–base reaction when an oxygen molecule acting as a Lewis base uses a pair of electrons to form a bond to an Fe^{2+} ion in the haemoglobin molecule. Once captured, the precious oxygen is transported in the blood stream to take part in other reactions deep inside our body.

Suffocation by carbon monoxide poisoning is another Lewis acid–base reaction. Now the carbon monoxide molecule, CO, can usurp the place of oxygen and attach Lewis-like to the Fe^{2+} ions in haemoglobin. This attachment is stronger than oxygen is able to achieve, so the usurper blocks the attachment of oxygen, and there is none transported to where it is needed and the victim suffocates. This is suffocation at a molecular level, not just the blocking of an airway. The poisoning effect of the cyanide ion, CN^-, I mentioned earlier is similar, but it blocks a cascade of electron transfer reactions later in the respiration process.

The subtlety of organic chemistry

Organic chemists are magicians, or general officers commanding, when it comes to deploying these fundamental types of reactions. They need to be, because the molecules they aspire to build are often delicate traceries of atoms, and one atom out of place could render a pharmaceutical inactive or set back research for months. Over the decades of development of organic chemistry, chemists have accumulated a fund of experience in knowing how to coax atoms into the appropriate arrangement to suit their need, sometimes in sequences of reactions with dozens of steps, any one of which might reduce a hard-won compound to the chemist's equivalent of rubble, a black, useless tar. The procedures often go by the names of the chemists who developed them. Computer software is also a help in devising strategies, just as it is used for establishing the workflow of a construction project.

The metaphor of a construction project can be taken further. Just as a partially completed component of the project might need to be protected while building goes on around it, so a partially constructed molecule might have tender regions that, without protection, would be centres of reaction and result in unwanted products. Thus, chemists sometimes might attach a small group of atoms to a region of the molecule either to shield a neighbouring region from attack or to conceal the atom to which it is attached. That protective group can later be stripped off, just like the protective shroud of a building.

I shall give just two examples of how organic chemists go about the business of building a molecule, perhaps one destined to be tested as a pharmaceutical, a dye, or an artificial flavour. Both are examples of a *substitution reaction*, in which an atom or group of atoms is substituted for one already present in a molecule. In each case the target atom is detected by the incoming reactant as a region of the molecule that has either a relatively thin or dense electron cloud. If the cloud is thin, the positively charged nuclei shine through, and a negatively charged reactant molecule will home in on it like a guided missile. Reactions of this nuclear-charge-detecting kind are called *nucleophilic substitutions*. If instead the cloud is dense, then the negative charges of the electrons will outweigh the positive charge of the nuclei, and an incoming missile that is positively charged will home in on the region. Such electron-rich-seeking reactions are called *electrophilic substitutions*.

When planning the construction of a molecule, a chemist needs to think about the way that the electron clouds are distributed in a molecule, and then choose the reactant molecule accordingly. They can be very subtle about the procedure, because it is possible to attach groups of atoms that suck electrons away from a region or alternatively push electrons on to it. By modifying the electron cloud in this way, a chemist can be reasonably confident that a reactant molecule will home in on the right atom and form a bond there.

I hope that, at this point, you are beginning to be able to sense the subtlety with which chemists go about the task of creating forms of matter that might not exist anywhere else in the universe. It will not be possible from this necessarily brief account to comprehend the details of how chemists contrive their reactions, but I hope that you will perceive the thoughtfulness behind their activities.

Chapter 5
Its techniques

Go into any modern chemical laboratory and you will find it a hybrid. An alchemist would recognize some of the apparatus; the rest would be wholly alien. There are only so many shapes for vessels to contain fluids and most of them have a clear ancestry in the past. But modern *analysis*, literally the breaking down of substances and in modern practice the identification of substances and the determination of their amounts and concentrations, makes use of sophisticated electronic and often automated equipment. Analysis is not the only pursuit in a laboratory, for its opposite, *synthesis*, literally the putting together but in practice the creation of desired forms of matter from simpler components, is a major component of a chemist's endeavour.

Classical laboratory equipment

I shall not dwell on beakers, flasks, and test-tubes, for their use for containing and mixing fluids is obvious. Some containers, though, are designed to deliver known amounts of liquids either to accord, as in a kitchen, with a precise recipe or as part of a method of quantitative measurement. An example of the latter is the use of a pipette (US: pipet; a little pipe) and a burette (US: buret; a word derived from the French word for a small vase or jug, although except to the imaginative it is nothing like one) in one of the classic procedures of chemistry, the titration of an acid with a base to determine the concentration of one or the other of them. (Why 'titration'? *Titre* is the French word for assay or test). The pipette is used to deliver a fixed amount of the basic solution to a conical flask; the burette is used to dribble in the acid until a colour change or an electronic signal from a detector signals that the base has been exactly neutralized. By noting the volume of acid added from the graduated burette and knowing its concentration, the concentration of the base can be determined.

One other class of apparatus is concerned with the separation of substances, perhaps to purify or isolate a product. One straightforward technique, when the product is a solid that has been precipitated when two solutions are mixed, is 'filtration': passing the resulting solution through a fine mesh. Another, often used when liquids need to be separated, is 'distillation': boiling the liquid mixture and condensing the vapour; the more volatile component of the mixture boils off first, and may be collected or discarded.

One highly sophisticated separation technique is 'chromatography'. This technique was born and named when it consisted of little more than noting that a drop of solution, perhaps taken from a flowering plant, would spread across absorbent paper and form bands of different colours that could then either be identified or collected. The name survives, but the technique has been immeasurably elaborated. Now, in a typical procedure, the sample to be analysed is passed through many metres of narrow tubing, the interior of which is coated with an absorbent solid. The components of the mixture stick to (the technical term is 'adsorb on') the surface to different degrees, and although they all make it to the end of the tube, they emerge at different times and can thus be collected separately and identified by other procedures. This technique is used to separate the myriad compounds that contribute to the flavour of a fruit and, in a more specialized form, to sniff out explosives, such as at security installations.

Spectroscopy

Much more interesting, and bewildering to the alchemist, is the electronic equipment in the room, showing only its screens and dials and not advertising immediately its purpose. Many of these procedures are forms of *spectroscopy*. The term is derived from the Latin word *spectrum*, or appearance, and the process of looking at the appearance; but 'looking' is now more sophisticated than visual inspection and 'appearance' far removed from its everyday meaning.

I shall begin with atomic spectroscopy. When an element is vaporized and heated, one or more of the electrons of an atom may be ejected from its normal distribution and briefly hang above the atom before collapsing back into its normal cloud again. That collapse gives an impulse to what we think of as the vacuum that surrounds the atom, and the impulse generates a pulse of light, a *photon*. The colour of the photon depends on the energy released in the collapse, with high-energy collapse giving a pulse of ultraviolet radiation and lower energy pulses giving visible light. The electrons of atoms can exist in a variety of energy states that are characteristic of the element, and as the electrons collapse from the state that they happen to have been promoted into they generate photons of the corresponding colours. We are all familiar with the yellow of street lighting, which is due to sodium atoms generating photons as they collapse into their normal state, and of red neon signs, which is due to the electron of a neon atom collapsing into its normal state. By noting the pattern of colours, by 'recording the spectrum', the element present can be identified.

The electrons of molecules behave in much the same way, but monitoring their possible energies is carried out in a somewhat different manner. Whereas the atomic spectroscopy that I have described makes use of the emission of light, molecular spectroscopy does the opposite: it makes use of the absorption of light.

Light that is passed through a sample can be thought of as a stream of photons. One of those photons will be absorbed if it collides with a molecule that can be excited into a higher energy state with a matching energy. The removal of such photons, their 'absorption', from the incoming stream will reduce the intensity of the beam, which will be recorded by a detector of some kind. To record the full absorption spectrum, the colour of the incident light is changed systematically, and the intensity that manages to survive passage through the sample is monitored. Because molecules have characteristic energy levels, their absorption spectra are unique, and can give a good indication of their identity.

I have focused on the absorption of photons by the excitation of electrons from their normal distribution in a molecule. That takes a lot of energy, and although many molecules absorb visible light (which is why the world is so colourful) the spectra I have described are commonly observed by using ultraviolet radiation. Thus, the technique is known as 'UV-vis spectroscopy'. A closely related technique uses photons of infrared radiation, which have much lower energy than visible and ultraviolet photons. These photons can stimulate the vibrations of molecules, not their electron distributions. 'Infrared spectra' therefore show that vibrations can be stimulated. That is very helpful for analysing the groups of atoms present in a complex molecule because a CH_3 group, for instance, can waggle around with one energy and a CO group can waggle around with a different energy.

Nuclear magnetic resonance

Perhaps the single most important analytical spectroscopic technique is 'nuclear magnetic resonance' (NMR). The word 'nuclear' raises a red flag wherever it occurs, which is why it has been dropped from the medical investigatory technique of 'magnetic resonance imaging' (MRI), a technique that is derived from NMR itself. Chemists on the whole are less squeamish than the public at large and retain the word 'nuclear', knowing that in this usage it has nothing at all to do with the perils of radioactivity.

The 'nuclear' of nuclear magnetic resonance refers to any nucleus of any atom, but I shall focus on its most common target, the proton, the nucleus of a hydrogen atom. A proton spins on its axis, like the Earth, and that spinning charge behaves like a tiny bar magnet. It can spin either clockwise or anticlockwise, and the corresponding little bar magnet has its North pole either up or down according to the direction of spin. When the spinning proton is in a magnetic field (in practice, an intense one generated by passing a current through a superconducting coil) the two orientations have different energies, and an incoming photon of the appropriate frequency can flip an upward pointing (low energy) proton into a downward pointing (high energy) proton. The matching of the photon frequency to the energy separation is the 'resonance' of the name; we do the same thing when we tune a radio to the frequency of a distant transmitter. When that takes place, the incoming stream of photons is attenuated, and the decrease in intensity is detected. The energy separation is not very great, and photons of radiofrequency radiation, just off the high-frequency end of an FM radio signal (100 MHz or so), are used.

It might seem a rather pointless activity to flip a proton from one orientation to another. The power of the technique—and that power cannot be underestimated—is that the precise frequency at which resonance occurs depends on where the proton, specifically the hydrogen atom of which it is the nucleus, lies in the molecule. Hydrogen atom nuclei with carbon atoms as neighbours resonate at different frequencies from those with oxygen or nitrogen atoms as neighbours, and so the spectrum of resonant absorptions, the 'NMR spectrum', portrays the neighbourhoods of all the hydrogen atoms in the molecule.

That's not all. The little bar magnets at the heart of hydrogen atoms in the same molecule interact with one another and modify each other's energies. That modification affects the resonant frequencies and gives rise to characteristic patterns of absorption, which is a huge help when trying to identify a molecule.

A carbon nucleus does not spin and so does not behave like a bar magnet and is invisible in NMR. That is a blessing, because otherwise even a quite simple organic molecule would give an impossibly complex NMR spectrum. However, carbon atoms can be revealed cautiously by replacing ordinary carbon atoms, carbon-12, with an isotope, carbon-13, which has an extra neutron in its nucleus and is magnetic. Judicious replacement of carbon-12 by carbon-13 can therefore be used to map the locations of carbon atoms too, and the identity and structure of the molecule can then be pinned down unambiguously.

Mass spectrometry

There is another totally different kind of spectrometer that does not use absorption or emission and gives an entirely different insight into the identity of a molecule. In a 'mass spectrometer' a molecule is blasted apart and its fragments weighed, the composition of the molecule then being inferred from the masses of the fragments.

The molecular shattering is carried out with a blast of electrons, which strike the molecule, distort the electron clouds holding it together, and give rise to a number of electrically charged fragments. These charged fragments are accelerated by an electric field and pass between the poles of a powerful magnet, which bends their paths to an extent that depends on their mass and the strength of the field. Fragments of a particular mass will fall on a detector and give a signal. As the magnetic field is changed, fragments of different masses will come into view by the detector, and the spectrum, now a 'mass spectrum' of masses of fragments, is interpreted in terms of the structure of the parent molecule rather as a smashed vase can be rebuilt from its fragments.

X-ray diffraction

In biology, structure is crucial to function. Structure is almost everything in chemistry, and especially so where chemistry merges with biology and chemists contribute to the study and elucidation of the action of the big protein molecules we know as enzymes. Although enzymes are hugely important for regulating the chemical reactions that constitute all aspects of life, they are not the only crucial components of organisms (which include humans). Inheritance is enabled by DNA, scaffolding by rigid proteins and bone, and perception and thought by molecules that detect and convey messages. The mechanism of the whole body of an organism is modulated by its marinating molecules.

One of the most powerful tools for discovering structure is 'X-ray diffraction' or, because it is always applied to crystals of the substance of interest, 'X-ray crystallography'. The technique has been a gushing fountain of Nobel prizes, starting with Wilhelm Röntgen's discovery of X-rays (awarded in 1901, the first physics prize), then William and his son Laurence Bragg in 1915, Peter Debye in 1936, and continuing with Dorothy Hodgkin (1964), and culminating with Maurice Wilkins (but not Rosalind Franklin) in 1962, which provided the foundation of James Watson's and Francis Crick's formulation of the double-helix structure of DNA, with all its huge implications for understanding inheritance, tackling disease, and capturing criminals (a prize shared with Wilkins in 1962). If there is one technique that is responsible for blending biology into chemistry, then this is it. Another striking feature of this list is that the prize has been awarded in all three scientific categories: chemistry, physics, and physiology and medicine, such is the range of the technique and the illumination it has brought.

To understand the basis of the technique it is essential to know that X-rays are beams of very short wavelength electromagnetic radiation, like light, but with wavelengths a thousand times shorter (around 100 pm (picometres), about the diameter of an atom). The second essential piece of information is that X-rays, like all waves, interfere with each other: where peaks coincide they are brighter; where peaks coincide with troughs, they are dimmer. When an object is put in the path of a beam of X-rays it scatters them, and scattering by different parts of a molecule results in beams that travel to a detector by different paths and so can interfere with each other in a variety of ways. This interference caused by an object in its path is the 'diffraction' part of the name.

In an X-ray diffraction determination, a tiny crystalline sample is rotated in the path of a beam of X-rays, and a detector is made to travel all over a surrounding sphere of locations, detecting the glints of constructive interference as it goes. From that huge number of observations a mathematical trick can be used to establish the arrangement of atoms in the sample. The technique is now extensively automated, with an integrated computer controlling the collection and interpretation of data.

The most challenging part of the determination is making the crystal that is essential to the technique, especially for the big molecules that are one of its principal targets of study. If, however, it is only identification of a substance, such as a mineral, that is required, then it is possible to use a simpler technique in which the sample is a powder spread on a plate. When a beam of X-rays is directed on to the powder, the resulting 'powder diffraction pattern' is characteristic of the substance, and it can be identified by referring to a library of patterns.

Why X-rays? Diffraction patterns are obtained when the wavelength of the radiation is comparable to the scale of the structures that cause it, in our case the atoms, causing the diffraction. It happens that X-ray wavelengths are comparable to atom–atom separations in molecules, so are ideal for the purpose.

Picturing surfaces

The interior of solids is a fascinating place, but the action often takes place on the surface. For instance, in catalysis, the acceleration of reactions by the presence of an otherwise non-participating substance, often occurs through a mechanism involving the attachment of reactants to a solid surface, sometimes ripped apart, where they are rendered ripe for reaction with other reactants. The chemical industry owes its existence to catalysts, so the study of events at surfaces is of great importance.

Surfaces, despite being the outward show of a solid, were hard to study until a few years ago when a dramatic new technique burst on to the scene. This technique is so sensitive that it can portray the individual atoms on the surface and also molecules stuck to it. It comes in two forms: 'scanning tunnelling microscopy' (STM) and 'atomic force microscopy' (AFM).

At first sight, STM might seem rather unlikely. A needle is pulled out to form a very fine point, and then swept in successive rows across the surface being studied. The flow of electric current between the needle point and the surface is monitored and mapped on a screen: atoms that protrude from the surface lie close to the passing needle point and give rise to a surge in current, which is portrayed as a peak on the screen. The success of the method depends on a quantum mechanical effect called 'tunnelling' (hence the tunnelling in the name of the technique), in which electrons are able to cross forbidden regions, in this case the gap between the surface and the needle. Tunnelling is very sensitive to the width of the gap, so the scan across the surface can pick up atom-size variations in the surface itself, and also show up, and depict in compelling detail, the shapes of molecules stuck to the surface.

It is widely claimed that atoms are too small to be seen, however, provided we enlarge our vision of 'seeing' to the graphical portrayal of the variation in tunnelling current, then STM denies that claim and provides us with the most extraordinarily compelling images of individual atoms and molecules. Even the clean surfaces themselves are compelling, with Mars-like mountains and cliffs where atoms pile high and chasms where they have been lost. Surfaces now are open to detailed, direct inspection.

Atomic force microscopy brings direct action to surfaces. Instead of passively observing, the tip is used to move atoms around on the surface by nudging them from place to place. Apart from allowing such entertainments as 'nanosoccer' by moving a C_{60} 'buckyball' over the surface, exquisite control can be exercised on the arrangement of individual atoms. If the tip is coated with molecules of a certain kind, then by moving the tip patterns can be written on to the surface and structures built up on a nanoscale (see Chapter 7).

Computational chemistry

An instrument that has transformed chemistry, just as it has transformed life in general, in recent decades is the computer. Just about all laboratory procedures, except the most primitive, are controlled by computers. As we have just seen, computers are intrinsic to X-ray crystallography, and are essential to the interpretation of diffraction patterns. They are also essential to modern NMR, where special techniques are used to observe the spectrum and need extensive mathematical manipulation to mine for the actual spectrum. There is, however, an application of computers in their own right, that of the computation and graphical portrayal of molecular structures. This is the field of 'computational chemistry'.

Together with weather forecasters and code breakers, chemists are among the most demanding users of powerful computers, although such is progress in computational hardware that much analysis can now also be carried out on a tablet or even a smartphone.

One wing of computational chemistry takes the subject all the way back to the quantum mechanical description of the distribution of electron clouds in molecules, and sets out to calculate those distributions. Such calculations involve a great deal of numerical manipulation and a variety of approximations. Although the output is essentially just a list of numbers relating to the density of the electron cloud throughout the molecule, those numbers are brought to life and rendered digestible by graphical displays of the cloud that enable chemists to assess the likely behaviour of the molecule. One very important application, which is greatly helped by the ability to picture the regions of dense and sparse electron clouds, is in the assessment of the pharmaceutical activity of a molecule and the initial screening of likely pharmacologically active compounds before they are tested *in vivo* on animals.

The second, closely related wing of computational chemistry has to do with how a protein folds into its active shape. A protein molecule is just a long chain of chemically linked molecules (amino acids), yet it folds into helices and sheets and they in turn fold into a reasonably rigid structure that is essential to its function. Although the forces that act between different parts of the same molecule are well understood, it is still an elusive problem to see how all those varied forces conspire to twist the chain into its final shape. Nature does it: we do not yet understand how. As a part of the attack on the problem, computers are used to track the links in the molecular chain as they wriggle and writhe into their final shape in an attempt to understand how Nature does it without so much as a thought.

Computers are used to study the behaviour of little molecules too. One molecule is sent flying towards another (in the imagination of the software), and calculations are used to watch what happens in the most intimate moments of a chemical reaction, when molecules collide, old bonds weaken, and new bonds form.

Modern approaches to synthesis

Now I would like to switch attention away from investigation towards synthesis and to a particular version of synthesis that is currently in vogue, where chemists do not always have a clue about what it is that they have made. I speak of 'combinatorial chemistry'.

The traditional procedure for making compounds is to work on one at a time, with a target clearly in mind. In combinatorial chemistry hundreds, even thousands, are made simultaneously and then examined for the appropriate behaviour, sometimes in similar herds, and when promising candidates have become apparent only they are analysed, their identities determined, and then used as the basis for further work.

The procedure originated with the synthesis of short strips of protein-like molecules known as peptides, and I shall use them to illustrate how it is done. The same strategy has been developed for a range of other types of compound, and has provided an important boost to the throughput of 'drug discovery', the process of formulating medicinally active compounds.

Suppose we have three amino acids, A, B, and C. In Round 1 we prepare a container with A in it, and proceed to carry out a reaction with all three acids, which will result in the compounds AA, AB, and AC. We then bring these three compounds together, mix them up, and divide them into three equal portions, each one containing all three compounds. That is the end of Round 1. Round 2 repeats the process, and in one container, the one that reacts with A, we shall get the three compounds AAA, ABA, and ACA, in the second container, which reacts with B, we get AAB, ABB, and ACB, and similarly for the third container, where we make ACA, ACB, and ACC. These nine compounds are used as the starting point for the next round. In practice, all 20 naturally occurring amino acids might be used rather than the miserly three of this illustration, and four successive rounds give 400, 8,000, 160,000, and 5,200,000 compounds. Thus, millions of compounds can be made (by robots) in almost the twinkling of an eye.

Thus it is that chemists make substances, not always knowing or caring what they have made, and hoping that amid the plethora of products there lie pearls. Clearly, there is a considerable bookkeeping task needed to keep track of the possible substances in each mixture (for instance, in our three-amino-acid example, after the second round the first container has only three candidates present), and a computer keeps track of what the robots are up to. If in some subsequent test the contents of a mixing vessel show a certain biological activity, such as inhibiting a malfunctioning enzyme that is responsible for a disease, then those contents can be candidates for separation and identification, the rest being washed away as of no interest.

There was a stage, only a few years ago, when chemists were proud to have made and identified about 10,000,000 compounds. Now they might make several times that number in a month, and only occasionally bother to determine what they have made. Such is progress.

Chapter 6
Its achievements

I have already remarked that life without chemistry would see us back in the Stone Age. Almost all the infrastructure and comforts of the modern world have emerged from chemical research. In the primitive days of the subject, when curiosity, tradition, and alchemy were collaborators and science was still the merest green shoot, that research was unaided by correct theory and progress was haltingly slow. Now that the subject is mature, with curiosity in fruitful alliance with understanding and exploitation, the research is largely rational and its achievements substantial.

In the broadest possible terms, chemists have discovered how to take one form of matter and conjure from it a different form. In some cases they have discovered how to take raw material from the Earth, such as oil or ore, and to produce materials directly from them, such as petroleum fuels and iron for steel. They have also discovered how to harvest the skies, how to take the nitrogen of the atmosphere and convert it into fertilizer. They have also discovered how to make highly sophisticated forms of matter suitable for use as fabrics or as the substances needed for what currently we regard as high technology, knowing that there is still higher technology to come and which will be enabled, we can be confident, by chemistry.

Earth, air, fire, and water

I shall begin this account of chemistry's numerous achievements by considering the famous four so-called elements of antiquity: earth, air, fire, and water.

First water, the absolutely essential enabler of life both at the level of individual organisms and at the level of global societies. Chemistry has made communal living possible through its use to purify water and rid it of pathogens. Chlorine is the principal agent for enabling cities to exist: without it, disease would be rampant and urban living a gamble and more akin to urban dying, just as it used to be. Chemists have found ways of extracting this element from an abundant source—sodium chloride, common salt—by using electrolysis to oxidize the chloride ions in the molten salt and by stripping off an electron from each one, so converting them to the element. The virile gas chlorine then goes on to attack the pathogens, rendering them harmless.

Chemists are at the forefront of the battle to obtain potable water from brackish water, from poisoned water in aquifers, and from that most abundant source of all, the oceans. They have contributed in direct ways to this crucial task by developing 'reverse osmosis', the process in which water is squeezed through membranes that filter out the ions that render it undrinkable. They contribute too in indirect ways by developing the membranes that can withstand the high pressures involved and contribute to the efficiency of the process. It goes without saying that chemists' traditional skills of analysis, discovering what is present, what can be tolerated, and what it is essential to remove, are crucial to this endeavour.

Then take earth, food's source. As the global population grows and the productive land area is eroded, so it becomes more and more important to coax crops into greater abundance and fecundity. Genetic engineering (a chemical technique, performed literally in fruitful collaboration with biology) is one way to proceed, but remains controversial for a variety of reasons, some plausible, others not. The traditional way to encourage abundance is to apply fertilizers. Here, chemists have contributed substantially by finding economically viable sources of nitrogen and phosphorus and ensuring that they can be converted to a form that can be assimilated by plants.

Air supplies the earth. Nitrogen (N), one of the elements essential to agriculture, is astonishingly abundant, making up nearly three quarters of the atmosphere; but it is there in a form that cannot be assimilated by most plants. This stubborn inertness is due almost entirely to the fact that the two nitrogen atoms of a nitrogen molecule (N_2) are strapped together by a powerful triple bond, three shared pairs of electrons, and are notoriously difficult to separate. Indeed, that is the principal reason why atmospheric nitrogen is so abundant in the air: it simply remains aloof from most attempts to react with it, requiring lightning bolts or the bacteria associated with certain leguminous plants to trap it chemically.

One of chemistry's greatest achievements, made in the opening years of the 20th century under the impetus not of a humane desire to feed but of an inhumane desire to kill, was to discover how to harvest nitrogen from the air and turn it into a form that could be absorbed by crops (and used to make explosives). This achievement, by Fritz Haber and Carl Bosch, was a landmark in the chemical industry, for as well as depending on the discovery of appropriate catalysts to facilitate the reaction between nitrogen and hydrogen gases to form ammonia (NH_3), it required the development of industrial plant that operated at temperatures and pressures never previously attained. But the process, which is still used globally today, remains energy intensive. It would be wonderful if the processes known to occur in the bacteria that inhabit the nodules of the roots of alfalfa, clover, beans, peas, and other legumes could be emulated on an industrial scale and harnessed to harvest atmospheric nitrogen. Chemists have put decades of research into this possibility, dissecting in detail the enzymes that bacteria use in their quiet and energy-efficient, low pressure, low temperature way. There are glimmerings of success, but no method is yet commercially viable.

Phosphorus (P) is abundant too, being the remains of prehistoric animals. Their bones of calcium phosphate and their special internal power source, the molecules of ATP (adenosine triphosphate) that power every one of our and their cells, lie in great compressed heaps below the oceans of the world as phosphate rock. Here chemists help to mine the dead to feed the living, for they find ways to extract the phosphorus from these buried sources and use it again in the great cycle of sustainability.

After water, air, and the food that springs from the earth, we need energy, the representative of fire in this quartet. Nothing happens in the world without energy, and civilizations would collapse if it ceased to be available. Civilizations advance by deploying energy in ever greater quantities, and chemists contribute at all levels and in all aspects of the development of new sources and more efficient applications of current sources.

Petroleum is, of course, an extraordinarily convenient source of energy, as it can be transported easily, even in weight-sensitive aircraft. Chemists have long contributed to the refinement of the raw material squeezed and pumped from the ground. They have developed processes and catalysts that have taken the molecules provided by Nature and used them to cut the molecules into more volatile fragments and reshape them so that they burn more efficiently. But burning Nature's underground bounty might by future generations be seen as the wanton destruction of an invaluable resource, akin to species extinction. It is also finite, and although economically viable new sources of petroleum are constantly, for the time being at least, being discovered, it is proving hazardous and increasingly expensive to extract it. We have to accept that although an empty Earth is decades off, one day it will arrive and needs to be anticipated.

Where do chemists currently look for new sources? The Sun, that distant, furious, nuclear fusion furnace in the sky, is an obvious source, and the capture of its energy that Nature has adopted, namely photosynthesis, is an obvious model to try to emulate. Chemists have already developed moderately efficient photovoltaic materials, and continue to develop their efficiency. Nature, with her 4 billion year start on laboratory chemists, has already developed a highly efficient system based on chlorophyll, and although the principal features of the process are understood, a challenge for chemists is to take Nature's model and adapt it to an industrial scale. One route is to use sunlight to split water into its component elements, desirable hydrogen and already abundant oxygen, and to pipe or pump the hydrogen to where it can be burned.

I say 'burned'. Chemists know that there are more subtle and efficient ways of using the energy that hydrogen and hydrocarbons represent than igniting them, capturing the energy released as heat, and using that heat in a mechanical, inefficient engine or electrical generator. *Electrochemistry*, the use of chemical reactions to generate electricity and the use of electricity to bring about chemical change, is potentially of huge importance to the world. Chemists have already helped to produce the mobile sources, the batteries, that drive our small portable devices, such as lamps, music players, laptops, telephones, monitoring devices of all kinds, and increasingly our vehicles.

Chemists are deeply involved in collaboration with engineers in the development of 'fuel cells' on all scales, from driving laptops to powering entire homes and conceivably villages. In a fuel cell, electricity is generated by allowing chemical reactions to dump and extract electrons into and from conducting surfaces while fuel, either hydrogen or hydrocarbons, is supplied from outside. The viability of a fuel cell depends crucially on the nature of the surfaces where the reactions take place and the medium in which they are immersed.

Even nuclear power, both fission and one day fusion (the emulation on Earth of the Sun), depend on the skills of chemists. The construction of nuclear reactors depends on the availability of new materials, and the extraction of nuclear fuel in the form of uranium and its oxides from its ores involves chemistry. Everyone knows that one fear that holds back the development and public acceptance of nuclear energy, apart from political and economic problems, is the problem of how to dispose of the highly radioactive spent fuel. Chemists contribute by finding ways to extract useful isotopes from nuclear waste and by finding ways to ensure that it does not enter the environment and become a hazard for centuries.

Artefacts from oil

I have alluded to the seemingly wanton destruction of an invaluable resource when the complex organic mixture we know as oil is sucked from the ground where it has lain for millennia and then casually burned. Of course, not all the oil is used in engines and its combustion products spewed out through the exhausts of our cars, trucks, trains, and aircraft. Much is extracted and used as the head of an awesome chain of reactions that chemists have developed and which constitute the petrochemical industry.

Look around you and identify what chemists have achieved by taking the black, viscous crude oil that emerges from the Earth, subjecting it to the reactions that they have developed, and passing on the products to the manufacturers of the artefacts of the modern world.

Perhaps the greatest impact of these processes has been the development of plastics. A century ago the everyday world was metallic, ceramic, or natural, with objects built from wood, wool, cotton, and silk. Today, an abundance of objects are built from synthetics derived from oil. Our fabrics have been spun from materials developed by chemists, we travel carting bags and cases formed from synthetics; our electronic equipment, our televisions, telephones, and laptops are all moulded from synthetics. Our vehicles are increasingly fabricated from synthetics. Even the look and feel of the world is now different from what it was a hundred years ago: touch an object today, and its texture is typically that of a synthetic material. For this transformation, we are in debt to chemists who have discovered how to chop up the long molecules exuded by the Earth and then reassemble them into very long chains in the process of polymerization. Thus, ethylene ($CH_2=CHX$, with $X = H$) is spun into polyethylene and used for everything from shopping (as plastic bags, a mixed blessing) to helping win World War II (as cladding for radar cables). As I mentioned in Chapter 4, when X is chlorine, Cl, the monomers are spun into PVC, which has taken over from wood and metal in much construction work.

Although the convenience of plastic bags is perhaps outweighed by their blight on the environment, think of what we would not have if we had none of the polymeric materials invented by chemists and then fabricated in bulk. Think of a world without nylon and the polyesters of fabrics for clothing, upholstery, and decoration. Think of a world with only heavy metal containers for drinks, food, and household fluids. Think of a world without all the little plastic artefacts of everyday life, switches, plugs, sockets, toys, knife handles, keyboards, buttons...the list is almost literally endless, so ubiquitous is the presence of chemistry-generated polymeric materials.

Even if you mourn the passing of many natural materials, you can still thank chemists for their preservation where they are still employed. Natural matter rots, but chemists have developed materials that ensure that that decay is postponed. In short, chemists both provide new materials when those are judged appropriate or desirable, and provide means of prolonging the lives of natural materials when judgement and choice leads to their adoption.

Plastics are but one face of the revolution in materials that has characterized the last one hundred years and is continuing vigorously today. Chemists develop the ceramics that are beginning to replace the metals that we use in vehicles, so lightening them and increasing the efficiency of our transport systems with the consequent lessening of its impact on the environment. Ceramics, of course, are materials of great antiquity, for they are the stuff of pots (another largely unacknowledged contribution to the viability of social life). Modern ceramics are tailored more systematically from purified clay and other materials, and sometimes exhibit surprising properties. Who, for instance, would have suspected that one class of ceramics baked from an almost witch's brew of elements would have possessed the remarkable property of *superconductivity*, the ability to conduct electricity without resistance? This material, which operates at very low temperatures, but at much higher temperatures than the previously known superconducting materials and therefore more economically encouraging and acceptable, is still groping for applications, for fabricating wires and films from ceramics remains a challenging problem.

Ceramics include glass. Modern glass includes the optical fibre that constitutes the spine of our global communication system. Glass is fundamentally silica (silicon dioxide, SiO_2) from sand that has been purified, rendered molten, and then allowed to cool. Over the centuries chemists have fiddled with this fundamental composition and have given us the richly coloured 'stained' glass, where enthralling hues are caused by impurities added cautiously and selectively. Certainly in the early days the colours were developed by the skill and wisdom of glassmakers, then not specifically chemists. But it is now chemists who formulate the composition of glasses that in some cases are richly coloured but in others, for fibres in particular, are strikingly transparent and able to convey pulses of light over great distances with minimal attenuation.

The creation of colour

The world of human fabrication would be drab without the contributions of chemists. Vibrant colours were once the domain of the wealthy who could afford the expense of purchasing natural colours, such as Tyrian purple, extracted from the glandular mucus of certain sea-snails (*Bolinus brandaris*) where 12,000 snails are milked or wantonly squashed to derive little more than a gram of dye, barely enough to dye the hem of a cloak, or of lapis lazuli (the 'stone of heaven') from distant Afghanistan for deep appealing ultramarine. Then along came William Perkin (1838–1907) who, when attempting unsuccessfully to synthesize quinine, without the advantage of knowing its structure, in an aim to save the empire's armies and bureaucrats from malaria, stumbled instead on the dye he called mauveine, thereby saving sea-snails instead of soldiers from slaughter and incidentally founding the British chemical industry. Thus he laid the foundation for the generation of all his personal and much of Britain's national wealth.

Chemists have added a whole spectrum of colours to the material world, which is no longer drab, except when needed (as in camouflage), but instead can be anything from vibrantly assertive or demurely subtle. Not only is the range of colours now enormous, with fluorescence and reflective sparkle added to the range, but the colours are lightfast and can withstand the rigours of the laundry.

Chemically created colours are not confined to cloth. Pigments in general have been developed; not only the colouring materials themselves but also the support medium, as in the paints used in buildings and the acrylics used by artists. Think of the advances made in household paints, with improvements to their flow properties, their stability in aggressive atmospheres, and their range of colours, including colours that intentionally fade to show where paint is being applied.

Even the colours of television screens and computer monitors make use of solids that have been developed by chemists. Gone are the days of power-hungry, bulky cathode-ray tubes. Now we are in the world of liquid crystals, plasma displays, and OLEDs (organic light-emitting diodes). The liquid crystals and OLEDs are formed of molecules built by chemists that respond in special ways to electric fields and have made possible portable devices with visual displays.

The infrastructure of the everyday

Chemists are also responsible for developing the semiconductors that underlie the modern world of communication and computation. Indeed, one of the principal contributions of chemistry is currently the development of what could be regarded as the material infrastructure of the digital world. Chemists develop the semiconductors that lie at the heart of computation and the optical fibres that are increasingly replacing copper for the transmission of signals. The displays that act as interfaces with the human visual system are a result of the development of materials by chemists.

Currently chemists are developing molecular computers, in which switches and memories are based on changes in the shape of molecules. The successful development of such materials—and with the optimism so typical of science we can be confident that the endeavour will be successful—will result in an unprecedented increase in computational power and an astonishing compactness. If you are interested in the development of such smart materials, then you can expect to contribute to a revolution in computation. There is also the prospect of the development of quantum computing, which will depend on chemists being able to develop appropriate new materials and will result in an almost unforeseeable revolution in communication and computation.

Medicinal chemistry

I have barely mentioned health. One of the great contributions of chemistry to human civilization (and, it must be added, to the welfare of herds) has been the development of pharmaceuticals. Chemists can be justly proud of their contribution to the development of agents that fight disease. Perhaps their most welcome contribution has been the development of anaesthetics and the consequent amelioration of the prospect of pain. Think of undergoing an amputation 200 years ago, with only brandy and gritted teeth to sustain you! Next in importance has been the development, by chemists, often by observing Nature closely, of antibiotics. A century ago, bacterial infection was a deadly prospect, but now, through the availability of penicillin and its chemically modified descendants, it is curable. We have to hope that it remains that way, but we need to prepare for the opposite as bacteria evolve to evade their nemesis.

The pharmaceutical companies often come under attack for what many regard as their profligate profits and exploitation. But they deserve cautious sympathy. Their underlying motive is the admirable aim (albeit with an eye on profit) of reducing human suffering by developing drugs that combat disease. Chemists are at the heart of this endeavour. It is highly regrettable that the development is so expensive. Modern computational techniques are helping in the search for new lines of approach and helping to reduce reliance on *in vivo* animal testing, but extraordinary care needs to be exercised when introducing foreign materials into living human bodies, and years of costly research can suddenly be ruined if at the last stage of testing unacceptable consequences are discovered.

Closely allied with the contribution of chemists to the alleviation of disease is their involvement at a molecular level. Biology became chemistry half a century ago when the structure of DNA was discovered (in 1953). Molecular biology, which in large measure has sprung from that discovery, is chemistry applied to the functioning of organisms. Chemists, often disguised as molecular biologists, have opened the door to understanding life and its principal characteristic, inheritance, at a most fundamental level, and have thereby opened up great regions of the molecular world to rational investigation. They have also transformed forensic medicine, brought criminals to justice, and transformed anthropology.

The shift of chemistry's attention to the processes of life has come at a time when the traditional branches of chemistry—organic, inorganic, and physical—have reached a stage of considerable maturity and are ready to tackle the awesomely complex network of processes going on inside organisms: human bodies in particular. The approach to the treatment, more importantly the prevention, of disease has been put on a rational basis by the discoveries that chemists continue to make. If you plan to enter this field, then genomics and proteomics will turn out to be of crucial importance to your work. This is truly a region of chemistry where you can feel confident about standing on the shoulders of the giants who have preceded you and know that you are attacking disease at its roots.

Warfare, and other evils

Then there is the dark side of chemistry. It would be inappropriate in this account of chemistry's great achievements for no mention to be made of its ability to enhance humanity's ability to damage and kill, for those achievements have come at a cost, in some cases to human life, in others to the environment.

First, the advances made in killing and maiming. Chemists have been responsible for the development of gases for warfare and the optimization of explosives. Indeed, Fritz Haber, whom I have mentioned in connection with his invention of the process of the synthesis of ammonia that has led to the widespread availability of potent fertilizers, was also a leader in the development of poison gas. There is the hope that the elimination of such weapons will enable us to judge his net contribution to human life more kindly, despite how we judge his personality. Although governments have the responsibility for using such terrible weapons, the chemists who contributed to their development cannot, in my view, avoid our condemnation. No good has come from the development of chemical weapons that might be put in the opposite scale to mitigate our condemnation of them: they are pure evil. Numerous states, not all the most powerful but covering about 98 per cent of the world's population, have rejected them as illegal weapons of warfare, and it is to be hoped that the rest will follow suit and join the treaty banning them.

Chemical warfare can be waged by accident. Such was the case at Bhopal, India, in 1984, when the Union Carbide plant there ran out of control with the result, according to official sources, of nearly 4,000 deaths directly related to the disaster and a further 8,000 within two weeks, and with over 500,000 injured. Intentional chemical warfare has never been so successful. The proximate cause of the disaster was the entry of water into an over-stocked, under-cooled tank of the compound methyl isocyanate (CH_3NCO), an intermediate in the manufacturing process of a pesticide. The receding demand for the pesticide at the time had resulted in the accumulation of more than normal quantities of the intermediate. How the water entered remains disputed: the company maintains it was sabotage by a disgruntled employee; others maintain that it entered accidentally in a plant where the safety controls were disorganized, ineffectual, missing, inadequate, and disregarded. The ensuing reaction released 30 tonnes of toxic gas into the atmosphere, visiting death and incalculable physical and emotional suffering on the inhabitants of the densely populated surrounding shanty town.

Comments on the inherent dangers of chemical plants would be otiose, and suggesting that the risks outweigh the advantages would be banal. Only very rarely, however, do such catastrophes occur, and we have to hope that lessons learned from the awful price paid will instil better practice in design and operation of the plants that, in the main, contribute to our well-being.

The other dark facet of chemistry is its provision, improvement, and manufacture of explosives. Here the facet is not entirely black, for explosives are useful in quarrying and mining. The black facet is their use in bombs and in the provision of the impelling force of projectiles: bullets, mortars, and the like. Explosives are compounds that when detonated undergo a very fast reaction—essentially, the molecules fragment into tiny pieces that form a gas and the very fast generation of gas creates the destructive or impulsive shock of the explosion.

In the early days of explosives, gunpowder was king. Its action depends on the intimate intermingling of oxidizing agents (sulfur, potassium nitrate) and stuff that can be oxidized (charcoal, essentially an impure form of the element carbon). The migration of the electrons to the oxidizing agents from the carbon, dragging across atoms, results in a large number of little molecules, a gas. Since then, substances and mixtures have been developed that react more rapidly and accordingly give a sharper shock. Instead of the mingling of different components, chemists have worked towards the ultimate intimacy: ensuring that the oxidizing and oxidizable components are parts of the same molecule so that electron transfer and the ensuing atom rearrangement and molecular fragmentation are as fast as possible and that large numbers of small fragment molecules are formed to amplify the shock. Famous among such compounds is nitroglycerin. This highly unstable compound was tamed when Alfred Nobel (1833–96) discovered that it could be absorbed into a type of porous clay, so forming dynamite and in due course providing the funds for the establishment of one of the greatest conscience-appeasing foundations, the Nobel Foundation, committed as its prizes are to the enhancement of the human condition and the propagation of peace.

Environmental issues

While we are in this embarrassingly negative corner of chemistry, I cannot avoid that other great pointed finger, the one directed at the environmental damage laid at the subject's door, or at least at its drains. It is impossible to deny that the unwanted effluent of the chemical plant has wrought ecological havoc. Ever since Perkin's factories turned the nearby canals red, green, and yellow according to the manufacturing priorities of the day, mankind's aspiration for its own betterment has been at an environmental cost. In fact, the green shoots of environmental pollution, if that is not too ironical a term, can be traced back to the Greeks and Romans, for analysis of ice cores laid down in those eras show traces of the consequences of metal working.

The way forward is either legal or chemical. The legal constrains by the prospect of punishment; the chemical avoids by elimination at source. The latter, always the better mode of action, depends on developments of chemistry itself and has inspired the politico-environmento-chemical movement of *green chemistry*. In broad terms, green chemistry aims to minimize the impact of chemical manufacturing processes on the environment by strict guidelines about the use of materials and the elimination of waste.

The protagonists of green chemistry begin with the plausible proposition that it is better to prevent waste than to clean it up after it has been generated. The implication of that fundamental principle is that whatever is used as starting materials in a process should appear in (as close as it is possible to) its entirety in the final product: whatever atoms go in should all appear in the molecules of the product, with as few as possible discarded as unwanted. It is in this implication that there are considerable economic and technological impacts, and therefore commercial reluctance, for processes and plants need to be designed accordingly and specific raw materials acquired from inconvenient sources, possibly at great expense.

With the process optimized, and particularly if the optimization is beyond technological and economic grasp, the procedures should be designed to avoid or at least minimize the involvement, not just as waste but also as potentially escapable intermediates, of toxic compounds. That requirement is also required of the final product, which should offer minimum risk of toxicity for human life (as the formalizers of the principles identified, but it seems more than appropriate to add organisms in general) and the environment. The restraint also applies to auxiliary materials employed in the process, particularly the liquids that are used as solvents and might, perhaps 'might' for some current processes becomes 'must', be released into the environment, even in small amounts, as leaks develop in the recycling procedures. Chemists, even for their own miniscule laboratory procedures, are essential sniffers out of benign solvents and the development of reactions that take place in these unfamiliar novel environments.

Another ideal aspiration of the proponents of green chemistry is that the feedstock should be renewable. Renewability can take a variety of forms, but all avoid the gouging out of resources from the Earth. Nature furnishes crops each year, and they count as renewable due to the benevolence of the Sun and its powering of the recycling of carbon dioxide through the medium of photosynthesis. Materials other than carbon dioxide can be recycled and plans have been proposed for the treatment of landfill as mines, but that resource is hazardous and not open to geological judgement.

The proponents of green chemistry recognize another contribution to waste and pollution: the role of energy in a chemical process. All requirements of energy make demands on the environment, either through the requirement of fuel or the impact of the exhaust on the atmosphere. Ideally, all procedures should take place without the need to heat and, even more expensively and destructively, cool.

Then there are a number of more technical requirements for the process to be as green as possible. Many procedures in organic chemistry, as in the fabrication of pharmaceuticals, require intermediate steps in which molecules are modified temporarily on their way to becoming the final product. Each step needs special conditions, its own reagents, and perhaps a variety of noxious solvents. The procedure shifts towards the green end of the production spectrum by minimizing these intermediates and looking for more direct routes from feedstock to product.

Bright green chemists look beyond the process itself to the whole lifetime span of its product and look for ways to ensure that at the end of its functional lifetime the product and anything into which it decays will not be toxic or degrade while in the environment into toxic remnants. The 'whole lifetime' consideration includes the anticipation of disaster during the manufacturing process itself (recalling Bhopal), with the precautionary implication that whatever is produced or stored should, in the event of accident, have a minimal effect on the environment. The mitigation of the possibility of catastrophe entails the ceaseless and reliable analysis of all the components and conditions of reaction and storage vessels and fail-safe monitoring procedures that cannot, as at Bhopal, be ignored or circumvented.

Such are the aspirations of green chemistry. The underlying consideration is that it is essential to appeal to chemistry to solve, and preferably avoid, the problems it might cause. There is always, of course, a tension between commercial profit and social and environmental responsibility, this tension not being helped by low levels of supervision in some environments which allows industry to get away, almost literally, with murder.

Pandora's box has always been thus: meddling with Nature invariably entails risk. Chemists meddle at the very roots of material Nature, taking the atoms she provides and recasting them into compounds that are alien to her and which, intruding into her ecosystem, can upset the fine balances of life. With this Merlin-like ability to conjure with atoms come responsibilities, which have not always been recognized in the past, but under social pressures are now high in the chemical industry's awareness of its responsibilities.

The crucial consideration, however, is where reliable solutions to the world's problems will come from if it is not further development of chemistry. Chemistry holds the key to the enhancement of almost every aspect of our daily lives, from the cradle to the grave and all points in between. It has provided the material foundation of all our comforts, not only in health but in illness too, and there is no reason to suppose that it has reached its zenith. It contributes to our communications, both virtual and physical, for it provides the materials along which our electrons and photons travel in the complex network of patterns and interactions that result in computation. Moreover, it develops our fuels, rendering them more efficiently combustible and through catalysis minimizing their noxious products, and helps in the migration from fossil fuels to renewable sources, such as in the development of photovoltaic substances. Chemistry is the only solution to the problems it causes in the environment, be it in earth, air, or water.

The cultural contributions of chemistry

There is another achievement of chemistry that it would be inappropriate to ignore in this survey: that it gives insight into the workings of the material world, insights that range from rock to organism. Insights are an enhancement of the human condition, for they lend understanding to wonder and thereby add to our delight.

Through chemistry we understand the composition and structure of the minerals that constitute the landscape and can see into the structures of rocks and know why they are rigid, why they might glisten, why they might fracture and erode, and what they contain. We know why metals can be beaten into shape and drawn into wires, and through our knowledge of the arrangement of their atoms why some bend to our will but others snap. We understand the play that may be made with the properties of metals by forming alloys and steels. We understand the colours of gemstones and why we can see through glass but not through wood.

Through chemistry we can unravel and comprehend the once inscrutable mysteries of the natural world. We can understand the green of a leaf and the red of a rose, the fragrance of a herb and new-mown hay. We can understand, in a halting but increasing way, the intricate and complex reticulation of processes in the natural world that constitute the awesome and multifaceted property we know as life. We are beginning, even more haltingly, to understand the chemical processes in our brains that enable us to perceive, wonder, and understand.

Although chemistry does not deal with the ultimate fabric of the material world, the zoo of fundamental particles that lie in the domain of fundamental particle physics, it deals with combinations of them, atoms, that have distinguishing personalities. Through chemistry we have come to understand the personalities of the elements, understanding why they have these personalities through the structures of their atoms and why they enter into certain combinations but not others. Through chemistry, the very stuff of chemistry, we know how to make use of these personalities to build molecules and forms of matter that might not exist anywhere else in our galaxy.

We understand, through chemistry, the flavours of foods, the colours of fabrics, the texture of matter, the wetness of water, the changing colours of foliage in spring, summer, and autumn. Not every moment of our lives do we need to turn on understanding, for lying back in animal delight can be a pleasure of its own, just basking in the pleasure of our surroundings. But chemistry adds a depth to this delight, for when the mood moves us and the inclination impels, we can look beneath the superficial pleasures of the world and enjoy the knowledge that we know how things are.

Chapter 7
Its future

New elements go on being discovered, currently at the rate of one every year or so, meaning that the Periodic Table is getting bigger with more scope, in principle, for chemists to explore. Unfortunately, all these new elements are multiply useless: they are radioactive and so unstable that they vanish within fractions of a second. Moreover, no more than a few atoms of them are ever made, and immediately vanish in a puff of fundamental particles.

The edge of the unknown

There are theoretical reasons for suspecting that just a little further along in the Periodic Table, at the yet-to-be-made elements numbered about 126 (in 2013 we are up to 116, livermorium, with one or two others as yet unnamed and their sighting not yet confirmed but hinted at through the mists) that they will form what is known as an 'island of stability' and survive for significantly longer than those around them. It is unlikely, though, that any of them will have any useful applications except as test-beds for theories of nuclear structure. Chemists have no reason to think that they will provide an impetus to chemistry.

They have plenty of elements to get on with. New techniques are being developed that promise to extend the sensitivity, precision, and scope of observations. The ability to detect extraordinarily small quantities of materials is both a blessing and a curse. To understand the composition of a sample in exquisite detail brings understanding closer, to detect the hint that bombs have been in a terrorist's hands helps us to survive, but to find contaminants everywhere, for in this ever churning world that will ever be so, can just confuse and perhaps unnecessarily alarm.

New worlds

Important techniques that are being developed include those where small numbers of atoms and molecules can be studied rather than having to infer their behaviour from observations on bulk samples. Chemists want to know the intimacies of molecular interaction and transformation, and being able to examine the properties of molecules in isolation or as they come together and react, with bonds loosening, atoms shaking free, and falling into new arrangements, is the holy grail of chemistry (of physical chemistry, at least). For some years now it has been possible to watch molecules evolve on timescales of the order of femtoseconds (10^{-15} s, 1/1,000,000,000,000,000 s) and progress has been made to extend that scale to attoseconds (10^{-18} s, a thousand times shorter), when even electrons are frozen in motion and chemistry has finally become physics.

When we encounter tiny groups of atoms, interesting questions and special rules come into play. Take water, for instance: what is the smallest possible ice cube? It has been discovered that you need at least 275 water molecules in a cluster before it can show ice-like properties, with about 475 molecules before it becomes truly ice. That is a cube with about eight H_2O molecules along each edge. The importance of this kind of knowledge is that it helps us model the process of cloud formation in the atmosphere as well as understand how liquids freeze.

When dealing with tiny collections of atoms at low temperatures we have to accept that their behaviour is governed by quantum mechanics and that we should expect weird properties. All matter, including everyday matter, is also governed by quantum mechanics, but we deal with such vast numbers of atoms even in a pinch of salt that the weirdness is washed away and we perceive only averages, the familiarities of behaviour of common matter. These new states of matter that are starting to be made might have consequences of little importance for chemistry, but perhaps not: they might be perfect for storing data and for the development of quantum computing.

Chemists are contributing hugely to one emerging field where small numbers of molecules are present: the world of nanoscience and nanotechnology. Nanosystems (from the Greek *nanos*, a dwarf) are composed of entities about 100 nm (10^{-7} m, 1/10,000 of a millimetre) in diameter, and lie in the intermediate region between individual molecules (about a thousand times smaller) and bulk matter (about a thousand times larger). Frontiers are always fascinating places, and this notional frontier between the big and the small is no exception. The nanoparticles (notice how the prefix can be affixed to many nouns: there are more to come) are small enough for quantum effects to be relevant and for thermodynamics, once regarded as a finished theory, to be bewildered and in need of reconsideration.

Here is fruitful ground for physical chemists to explore, and to formulate and refine their conventional theories for application to these hitherto unconventional materials. Here too is where both organic and inorganic chemists have much to contribute, particularly in the fabrication of nanomaterials, for both organic and inorganic substances can be formulated to inhabit the nanoworld. Fabrication can be 'top-down', when nanostructures are carved out of macroscopic materials, like a sculptor at work on marble, or it can be 'bottom-up', when the nanostructures are built up brick by brick. The latter is particularly interesting, as the construction typically takes place by 'self-assembly'. In this hands-off procedure, molecules are constructed that, when shaken together, aggregate into the desired nanostructure, rather as we all might once have hoped that shaking a jigsaw puzzle would assemble the picture as the pieces interlocked spontaneously rather than going through the irksome business of linking them piece by piece by hand.

Nanotechnology, the development and application of nanomaterials, and nanoscience, their study in general, is currently all the rage in chemistry, and rightly so, for nanomaterials hold great promise. Whole institutes are being dedicated to their study. The potential applications of nanomaterials range across disciplines and are already central to many practical applications. For instance, they show superior light-harvesting characteristics compared to traditional silicon solar cells and have been incorporated into sensors for glucose in blood. Materials containing cadmium have been investigated extensively in the latter connection, with fears that the toxic element cadmium might be inappropriate to inject into human bodies; but recent results on primates seem to mitigate this fear. Nanorods, nanowires, nanofibres, nanowhiskers, nanobelts, and nanotubes have also been created, with potential applications in nanomachinery and nanocomputers.

Chemistry is preparing itself to play a major role in the miniaturization of computation. We have seen the impact of the reduction in size (and power consumption) between the early room-filling computers of the 1950s to the tiny, ubiquitous, powerful computers of today and their impact on society and daily life. That was a step from the scale of metres to centimetres, a hundredfold decrease in linear dimension and a millionfold decrease in volume and weight, scaling from room-size to pocket-size but accompanied by a huge increase in computational power. That decrease in size allied with an increase in capability and consequential increase in social impact can be repeated if current progress with the development of molecular computation bears fruit.

Computational procedures depend on two features: memory and manipulation. Memory is quite easy to achieve at a molecular level by causing a molecule to undergo a change of shape that is preserved and accessible to some kind of observation. For instance, a molecule might be caused to bend into a certain shape to represent 1 and bend into a different shape to represent 0. A variety of conformational changes are now available, such as a ring-like molecule sliding to either end of a rod-shaped molecule and staying there. Manipulation is more difficult, but comes down to achieving a certain output from a certain input. Chemistry, though, is all about outputs from inputs in the form of chemical reactions, including the output of light when two reagents meet.

Nature has already solved the problem of data storage in her development of DNA, and has evolved methods of extracting that information and turning it into organisms. Our memories are chemically encoded in as yet unknown ways in the brain and provide an immense but fragile and imperfectly stored database. DNA molecules have been used to perform simple arithmetical operations and to 'decide' the treatment necessary if they encounter a damaged protein molecule. Growing computers rather than making them is still science fiction, but there are hints of it on the horizon.

New dimensions

One remarkable recent development has been chemistry's migration from three to two dimensions. The common pencil-filling material graphite is a form of the element carbon in which the carbon atoms form flat sheets like chicken wire that, when impurities are present, slide over each other perhaps to be left as a mark on a page or to act as a lubricant. The individual sheets are called *graphene*, and the fact that they can be plucked off solid graphite by a very simple procedure helped to earn Andre Geim and Konstantin Novoselov the 2010 Nobel Prize (for physics).

Graphene itself is currently viewed as a great prize for physicists and potentially for engineers. It is one of the strongest materials known, with a breaking point 200 times greater than that of steel, yet is very light, weighing less than a gram per square metre. In the Nobel citation it is remarked that 'a 1 square metre hammock would support a 4 kg cat but would weigh only as much as one of the cat's whiskers'. Its extraordinary electronic, thermal, and optical properties are also of great interest, with among other potential applications the creation of loudspeakers with no moving parts and which can be moulded to different surfaces, and achieving in effect the room-temperature distillation of vodka, essentially filtering off the water.

Where do chemists stand confronting this two-dimensional crock of gold? It is currently being developed for laboratory techniques, such as its use as a sieve for separating molecules of different kinds (the production of biofuels is a target) and in desalination (the rendering of seawater potable). Although graphene itself does not readily adsorb gas molecules, its surface—it is almost entirely surface—can be chemically modified to be responsive to gases of different kinds, and their attachment modifies the electrical properties of the underlying graphene sheets so that their presence can be detected.

Chemists naturally wonder whether this two-dimensional wonderland can be inhabited by other materials, and whether those materials can circumvent some of the deficiencies of otherwise seemingly miraculous graphene. New materials of a graphene-like form have been made electrochemically, with compounds like molybdenum sulfide, tungsten sulfide, and more exotic materials based on titanium carbide. Some of these two-dimensional materials show semiconducting properties, which graphene lacks, and have already been fabricated into minute integrated circuits. Graphene itself is open to chemical modification, one procedure being to oxidize it to form graphene oxide. Flakes of this material aggregate into sheets of 'graphene paper' which, materials scientists are hoping, can form the basis of a whole new class of materials with tunable electrical, thermal, optical, and mechanical properties.

New applications

So vast are the applications of the new materials developed by chemists in collaboration with materials scientists, physicists, biologists, and engineers that I can do no more than stand in this Aladdin's cave of wonders and point around at random, knowing that I will miss a crucial development or example, but hoping to convey through just a few examples the impression that life is being transformed by this collaboration.

Thus, I point to self-cleaning glass. This labour-saving development is based on photochemistry and an understanding of the forces of attraction or repulsion between molecules, in particular the property that renders a surface 'hydrophobic', or water repelling. A typical self-cleaning glass is coated with a thin transparent layer of titanium dioxide, which responds to sunlight by breaking down chemically any dirt that happens to be deposited on it. The water-repelling surface means that any water, rainwater in particular, washes away the products of this photocatalysed decomposition without leaving dirty streaks.

I can point to smart fabrics. Smart fabrics can glow with different colours, perhaps representing the wearer's distribution of temperatures and, in a crude way, their emotional state. Or they can respond to the ambient conditions or the whim of the wearer by changing their appearance electrically. Not only must the fabrics be entertaining, they must also withstand the rigours of passage through the laundry and the stress of being worn, crumpled, and creased.

Catalysis is hugely important, as I have already indicated for industry, but is vital for the elimination of pollution from internal combustion engines. The catalytic converter now built into all our cars makes use of some highly sophisticated chemistry, for it must come into operation quickly as soon as the engine is started when it is cold (a significant proportion of pollution occurs then) yet continue to act when the engine is untouchably hot. Moreover, not only must the catalysts achieve reduction of nitrogen oxides to harmless nitrogen, they must also achieve the oxidation of carbon monoxide to carbon dioxide and complete the oxidation of unburned hydrocarbon fuel. Not only that, they also need to respond to the different conditions as the engine runs, such as the leanness or richness of the fuel/air mixture and sudden surges during acceleration. All this needs to be developed by chemists.

Perhaps nowhere is modern chemistry more important than in the development of new drugs to fight disease, ameliorate pain, and enhance the experience of life. *Genomics*, the identification of genes and their complex interplay in governing the production of proteins, is central to current and future advances in pharmacogenomics, the study of how genetic information modifies an individual's response to drugs and offering the prospect of personalized medicine, where a cocktail of drugs is tailored to an individual's genetic composition.

Even more elaborate than genomics is *proteomics*, the study of an organism's entire complement of proteins, the entities that lie at the workface of life and where most drugs act. Here computational chemistry is in essential alliance with medical chemistry, for if a protein implicated in a disease can be identified, and it is desired to terminate its action, then computer modelling of possible molecules that can invade and block its active site is the first step in rational drug discovery. This too is another route to the efficiencies and effectiveness of personalized medicine.

New discoveries

I do not want to give the impression that advances in chemistry are entirely confined to its applications. They are certainly headline-grabbing and affect us all. However, chemists are also engaged in the fundamental business of discovering more about matter and how it may be modified. Increasingly, they are becoming familiar with the workings of Nature at a molecular level, learning her ways, and stumbling on to features that might be astonishing and not have any immediate application except for that most precious of entities, knowledge. Fundamental research is absolutely vital to this endeavour, for it leads on to unforeseen discoveries, unforeseen understanding, and unforeseen applications of extraordinary brilliance.

In order to introduce a certain closing flourish, here I mention a single, singular, particular, purely academic recent discovery: Nature, chemists have discovered, can tie herself into knots. A class of molecules, it has been discovered to the researchers' astonishment and delight, can tie itself spontaneously into a trefoil knot. As a commentator (Fraser Stoddart) on this work remarked 'the new research illustrates some of the finest aspects of synthetic and physical organic chemistry and is one of these rare instances where stereochemistry is being expressed at its most elegant'.

Such is the joy, the intellectual pleasure, that modern chemistry inspires. I hope these pages have erased to some extent those memories that might have contaminated your vision of this extraordinary subject and that you have shared a little of that pleasure.

PERIODIC TABLE OF THE ELEMENTS

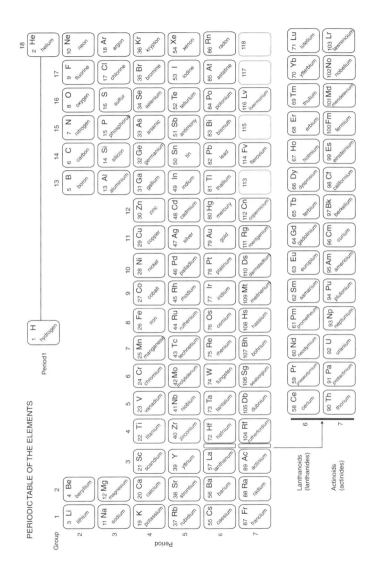

Glossary

Acid A proton donor (see *Lewis acid*).

Alkali A water-soluble base; a solution of a base in water.

Amino acid An organic compound of formula $NH_2CHRCOOH$ (R denotes a group of atoms, such as $-CH_3$, or something more complex).

Analysis The identification of substances and the determination of their amounts and concentrations.

Anion A negatively charged atom or group of atoms.

Atom The smallest particle of an element; an entity composed of a nucleus surrounded by electrons.

Base A proton acceptor (see *Lewis base*).

Bond A shared pair of electrons lying between two atoms.

Carbohydrate An organic compound of typical formula $(CH_2O)_n$.

Catalysis The acceleration of a chemical reaction by a species that undergoes no net change.

Cation A positively charged atom or group of atoms.

Chain reaction A reaction in which a molecule, ion, or radical attacks another, the product attacks another, and so on.

Complex A group of atoms consisting of a central metal atom to which are attached ligands.

Compound A specific combination of elements bonded together.

Diffraction Interference between waves caused by an object in their path.

Double bond Two shared pairs of electrons lying between two atoms.

Electrochemistry The use of chemical reactions to generate electricity and the use of electricity to bring about chemical change.

Electrolysis To achieve a chemical reaction by passing an electric current.

Electron A negatively charged subatomic particle.

Electrophile A species that is attracted to electron-dense (negative) regions.

Electrophilic substitution A substitution reaction in which one reactant is an electrophile.

Element A substance that cannot be broken down chemically into simpler substances; a substance composed of a single kind of atom. For a list of elements and their symbols, see the preceding Periodic Table.

Genomics The identification of genes and their complex interplay in governing the production of proteins.

Green chemistry The aim to minimize the impact of chemical manufacturing processes on the environment by strict guidelines about the use of materials and the elimination of waste.

Hydronium ion H_3O^+.

Hydroxide ion OH^-.

Intermediate See *Reaction intermediate*.

Ion An electrically charged atom or group of atoms (see *Cation* and *Anion*).

Isotopes Atoms with nuclei with the same atomic number (number of protons) but different numbers of neutrons.

Lewis acid An electron pair acceptor.

Lewis base An electron pair donor.

Lewis acid–base reaction A reaction of the form $A + :B \rightarrow A–B$ between a Lewis acid and a Lewis base.

Ligand A group of atoms attached to a central metal atom in a complex.

Lone pair A pair of electrons not involved directly in bond formation.

Mixture A mingling of substances without the formation of new chemical bonds.

Molecule The smallest particle of a compound; a discrete combination of atoms in a definite arrangement.

Monomer A small molecule used in a polymerization reaction.

Nucleophile A species that is attracted to electron-poor (positive) regions.

Nucleophilic substitution A substitution reaction in which one reactant is a nucleophile.

Oxidation The removal of electrons from a species; reaction with oxygen.

Photon A particle of electromagnetic radiation.

Polymer The product of a polymerization reaction.

Polymerization The linking together of small molecules to create long chains.

Product The material produced by a chemical reaction.

Protein A complex compound built from amino acids.

Proteomics The study of an organism's entire complement of proteins.

Proton The nucleus of a hydrogen atom.

Radical A species with at least one unpaired electron.

Reactant The starting material in a specified chemical reaction.

Reaction intermediate A species other than the reactants and products that is proposed to be involved in a reaction mechanism.

Reagent A substance used as a reactant in a variety of chemical reactions.

Redox reaction A reaction involving oxidation of one species and reduction of another; an electron transfer reaction.

Reduction The addition of electrons to a species.

Salt An ionic compound formed by the reaction of an acid and a base.

Solute A dissolved substance.

Species Used here to denote an atom, molecule, or ion.

Spectroscopy The observation of the absorption or emission of radiation by a sample.

Substitution reaction A reaction in which an atom or group of atoms is substituted for one already present in a molecule.

Superconductivity The ability to conduct electricity without resistance.

Synthesis The creation of substances from simpler components.

Titration The determination of the concentration of an acid (or base) by measuring the volume of an alkali (or acid) needed to neutralize it.

Transition metal A member of Groups 3 to 11 of the Periodic Table.

Further reading

For a survey of chemical thermodynamics, see my *Four Laws that Drive the Universe* (2007), reissued as *The Laws of Thermodynamics: A Very Short Introduction*, Oxford: Oxford University Press (2010).

For a broader introduction to the principles of chemistry, see my *Physical Chemistry: A Very Short Introduction*, Oxford: Oxford University Press (2014).

The variety of chemical reactions that is merely touched on in this volume is elaborated pictorially in my *Reactions: The Private Life of Atoms*, Oxford: Oxford University Press (2011).

For a broad survey of the principles and techniques of chemistry see my *Chemical Principles: The Quest for Insight*, with Loretta Jones and Leroy Laverman, New York: W.H. Freeman & Co (2013).

Others, of course, have written wonderfully and extensively on chemistry. The *Very Short Introductions* are particularly apposite, and include

Molecules, Philip Ball (2003)

The Elements, Philip Ball (2004)

The Periodic Table, Eric Scerri (2011)

For a survey of modern trends and applications of chemistry, see *The New Chemistry*, ed. Nina Hall, Cambridge: Cambridge University Press (2000).

Index

A

Absorption spectroscopy 56
Acetic acid 40
Acid 40
Activation barrier 34
Activation energy 33
Adenosine triphosphate 69
Adsorb 54
AFM 61
Air 67
Alchemy 1
Alfalfa 68
Alkali 40–1
Aluminium oxide 46
Amino acid 63
Ammonia 23, 35, 68
Amputation 76
Analysis 53
Analytical chemistry 10
Anion 20–1
Antibiotics 76
Atom transfer 46
Atomic force microscopy 61
Atomic number 15
Atomic spectroscopy 55
Atomic structure 14
Atoms 2

ATP 69
Attosecond timescale 87

B

Balance 1
Base 41
Basid reaction 45
Battery 46
Bhopal 78
Biochemistry 10
Biofuel 91
Biology as chemistry 5
Bolinus brandaris 74
Bond 24
Bond formation 19
Bosch, C. 36, 68
Bottom-up 88
Bragg, W. and L. 59
Breathing 50
Brønsted, J. 40
Burette 53–4
Burning 69

C

C_{60} 62
Carbon 8

Carbon dioxide 29
Carbon monoxide poisoning 50
Carbon valence 23
Carrot and cart 36
Catalysis 92
Catalyst 34–5
Catalytic converter 92
Cation 20
Ceramics 73
Chain reaction 48
Chemical equilibrium 35
Chemical kinetics 32
Chemical weapons 78
Chemistry, structure of 7
Chlorine 67
Chlorophyll 70
Chromatography 54
Classical mechanics 4
Clay 73
Cloud (of electrons) 17
Cloud layers 18
Combinatorial chemistry 64
Combustion 30
Common salt 21, 41, 67
Complex 49–50
Computational chemistry 62, 63
Cooking 34
Corrosion 43
Covalent bonding 22
Cyanide poisoning 50

D

Dalton, J. 2
Deuterium 16
Diffraction 60
Disease 76
Disorder 28
Distillation 54
DNA data storage 90
Double bond 24
Driving power 28
Drug development 93
Ductile 25

Dye 50
Dynamic equilibrium 36
Dynamite 79

E

Earth 66, 67
Electric conductivity 25
Electrochemistry 70
Electrode 46
Electrolysis 46
Electron 14–16, 43
Electron cloud 17
Electron pair 24
Electron sea 25
Electron shells 17–18
Electron spin 24
Electron transfer 43, 79
Electrophilic substitution 52
Element 2–3
Elements of antiquity 67
Elements, order of 15
Emission spectroscopy 55
Endothermic reaction 31
Energy 4–5, 27
Enthalpy 30
Entropy 28
Entropy increase 31
Enzyme 35
Equilibrium 35
Ethylene 72
Exothermic reaction 31
Explosives 78

F

Fabric 72
Faraday, M. 43
Femtosecond timescale 87
Fertilizer 66
Filtration 54
Fire 67
Fire-retardant 48
Fireflies 34

First Law of thermodynamics 27
Fission 71
Forensic chemistry 10
Franklin, R. 59
Free radical 47
Fuel cell 70
Fusion 71

G

Geim, A. 90
Genomics 93
Glass 73
Graphene 90
Graphical representation 63
Graphite 90
Green chemistry 11, 80
Gunpowder 79

H

Haber, F. 36, 68, 78
Haber–Bosch process 35
Haemoglobin 50
Health 76
Heat 30
Heavy hydrogen 16
Heavy water 16
Hodgkin, D. 59
Homeostasis 36
Hydrochloric acid 40
Hydronium ion 42
Hydrophobic 92
Hydroxide ion 41

I

Industrial chemistry 11
Infrared radiation 56
Infrared spectroscopy 56
Inorganic chemistry 9
Insight 83
Interference 60
Ion 20–1

Ionic bonding 22
Isotope 16

K

Knots 94

L

Laboratory equipment 39
Lapis lazuli 74
Lead-acid battery 46
Leguminous plants 68
Lewis acid 49
Lewis acid–base reaction 49
Lewis base 50
Life as titration 42
Lifetime span 82
Ligand 49
Liquid crystal 75
Lithium-ion battery 46
Livermorium 3
Lowry, T. 40
Lustrous 25

M

Macroscopic world 3
Magnesium burning 43–4
Magnesium chloride 44
Magnesium oxide 44
Malleable 25
Mass spectrometer 58
Mauveine 74
Mechanism of reaction 33
Medicinal chemistry 12
Mendeleev, D. 13
Metal 24–5
Metallic bonding 25
Methane 23
Methyl isocyanate 78
Microscopic world 3
Molecular biology 11–12, 77
Molecular computer 75

N

Nanobelt 89
Nanofibre 89
Nanomaterial 88, 89
Nanoparticle 88
Nanorod 89
Nanosoccer 62
Nanosystem 88
Nanotechnology 88
Nanowhisker 89
Nanowire 89
Neutron 15, 16
Newton, I. 4
Nitrogen 14, 66, 68
Nitrogen fixation
    68, 69
Nitroglycerin 79
NMR 56
NMR spectrum 57
Nobel Foundation 80
Nobel, A. 79
Non-metal 26
Novoselov, K. 90
Nuclear atom 14
Nuclear magnetic
    resonance 56
Nuclear spin 57
Nuclear waste 71
Nucleophilic substitution 52
Nucleus 14–15

O

Oil 71
OLED 75
Optical fibre 73
Ore reduction 44
Organic chemistry 8
Organic light-emitting diode 75
Organism 5
Organometallic chemistry 9
Oxidation 43, 44
Oxidation defined 44
Oxygen 14

P

Pathogens 67
Peptide 64
Periodic Table 13, 86
    structure of 18
Perkin, W. 74
Personalized medicine 93
Pesticide 78
Petrochemicals 69
Petroleum 69
Pharmaceutical companies 76
Phosphorus 14, 69
Photon 55, 56
Photosynthesis 47, 70
Physical chemistry 7–8
Pigments 50, 74
Pipette 53–4
Plastic bags 72
Plastics 48, 71
Poison gas 78
Polyester 72
Polyethylene 48, 72
Polymerization 48
Polystyrene 48
Polytetrafluoroethylene 48
Polythene 48
Polyvinyl chloride 48
Pots 73
Powder diffraction pattern 60
Preservation 72
Product 38, 39
Protection 51
Protein folding 63
Proteomics 93
Proton 15, 39–40
Proton spin 57
Proton transfer 39–40
PTFE 48
PVC 48, 72

Q

Quantum mechanics 4
Quinine 74

R

Radical 47–8
Radioactivity 16
Rate of reaction 32, 33
Reactant 38
Reaction mechanism 33
Redox reaction 45–6
Reduction 44, 45
Rontgen, W. 59
Rutherford, E. 15

S

Salt 41
Scanning tunnelling
    microscopy 61
Sea-snail 74
Seawater 91
Second Law of thermodynamics
    5, 28
Self-assembly 88
Self-cleaning glass 92
Semiconductors 75
Separation 54
Shaking and stirring 38
Silica 73
Single bond 24
Smart fabrics 92
Soap 40–1
Sodium 21
Sodium chloride 21,
    41, 67
Sodium sulfate 41
Solid-state chemistry 9
Solvents 81
Spectroscopy 8, 55
Spin 24
Stained glass 73
STM 61, 62
Subatomic particle 3
Substitution reaction 51
Sulfur 14
Sulfuric acid 40
Superconductivity 73

Surfaces 61
Synthesis 53

T

Teflon 48
Texture 72
Thermochemistry 31
Thermodynamics 4–5, 27
Thomson, J. J. 43
Titanium dioxide 92
Titration 53–4
Top-down 88
Toxicity 81
Transition metal complex 49
Triple bond 24
Tritium 16
Two-dimensional chemistry 91
Tyrian purple 74

U

Ultraviolet radiation 56
Union Carbide 78
UV-vis spectroscopy 56

V

Valence 29–30
Vibrational spectroscopy 56
Vitalism 9

W

Water as a base 42
Water as an acid 42
Water molecule 22, 29
Watson, J. 59
Weighing 2
Wilkins, M. 59

X

X-ray diffraction 59
X-rays 59–60

Index

Expand your collection of
VERY SHORT INTRODUCTIONS

1. Classics
2. Music
3. Buddhism
4. Literary Theory
5. Hinduism
6. Psychology
7. Islam
8. Politics
9. Theology
10. Archaeology
11. Judaism
12. Sociology
13. The Koran
14. The Bible
15. Social and Cultural Anthropology
16. History
17. Roman Britain
18. The Anglo-Saxon Age
19. Medieval Britain
20. The Tudors
21. Stuart Britain
22. Eighteenth-Century Britain
23. Nineteenth-Century Britain
24. Twentieth-Century Britain
25. Heidegger
26. Ancient Philosophy
27. Socrates
28. Marx
29. Logic
30. Descartes
31. Machiavelli
32. Aristotle
33. Hume
34. Nietzsche
35. Darwin
36. The European Union
37. Gandhi
38. Augustine
39. Intelligence
40. Jung
41. Buddha
42. Paul
43. Continental Philosophy
44. Galileo
45. Freud
46. Wittgenstein
47. Indian Philosophy
48. Rousseau
49. Hegel
50. Kant
51. Cosmology
52. Drugs
53. Russian Literature
54. The French Revolution
55. Philosophy
56. Barthes
57. Animal Rights
58. Kierkegaard
59. Russell
60. Shakespeare
61. Clausewitz
62. Schopenhauer
63. The Russian Revolution
64. Hobbes
65. World Music
66. Mathematics
67. Philosophy of Science
68. Cryptography
69. Quantum Theory
70. Spinoza
71. Choice Theory
72. Architecture
73. Poststructuralism
74. Postmodernism
75. Democracy
76. Empire
77. Fascism